Explainable AI Within the Digital Transformation and Cyber Physical Systems

Moamar Sayed-Mouchaweh
Editor

Explainable AI Within the Digital Transformation and Cyber Physical Systems

XAI Methods and Applications

 Springer

Editor
Moamar Sayed-Mouchaweh
Institute Mines-Telecom Lille Douai
Douai, France

ISBN 978-3-030-76408-1   ISBN 978-3-030-76409-8   (eBook)
https://doi.org/10.1007/978-3-030-76409-8

Preface

Explainable Artificial Intelligence (XAI) aims at producing explainable models that enable human users to understand and appropriately trust the obtained results. The produced explanations allow to reveal how the model functions, why it behaves that way in the past, present, and future, why certain actions were taken or must be taken, how certain goals can be achieved, how the system reacts to certain inputs or actions, what are the causes for the occurrence of a certain fault and how this occurrence can be avoided in the future, etc. The need for explanations is increasingly becoming necessary in multiple application domains, such as smart grids, autonomous cars, smart factory or industry 4.0, telemedicine and healthcare, etc., in particular within the context of digital transformation and cyber-physical systems.

This book gathers research contributions aiming at the development and/or the use of XAI techniques, in particular within the context of digital transformation and cyber-physical systems. This book aims to address the aforementioned challenges in different applications such as healthcare, finance, cybersecurity, and document summarization.

The discussed methods and techniques cover different kinds of designed explainable models (transparent models, model agnostic methods), evaluation layout and criteria (user-expertise level, expressive power, portability, computational complexity, accuracy, etc.), and major applications (energy, industry 4.0, critical systems, telemedicine, finance, e-government, etc.). The goal is to provide readers with an overview of advantages and limits of the generated explainable models in different application domains. This allows to highlight the benefits and requirements of using explainable models in different application domains and to provide guidance to readers to select the most adapted models to their specified problem and conditions.

Making machine learning-based AI explainable faces several challenges. Firstly, the explanations must be adapted to different stakeholders (end users, policy makers, industries, utilities, etc.) with different levels of technical knowledge (managers, engineers, technicians, etc.) in different application domains. Secondly, it is important to develop an evaluation framework and standards in order to measure the effectiveness of the provided explanations at the human and the technical levels. For instance, this evaluation framework must be able to verify that each explanation is consistent across similar predictions (similar observations) over time, is expressive in order to increase the user confidence (trust) in the decisions made, promote impartial and fair decisions, and improve the user task performance.

Finally, the editor is very grateful to all authors and reviewers for their valuable contribution. He would like also to acknowledge Mrs. Mary E. James for establishing the contract with Springer and supporting the editor in any organizational aspects. The editor hopes that this book will be a useful basis for further fruitful investigations for researchers and engineers as well as a motivation and inspiration for newcomers in order to address the challenges related to energy transition.

Douai, France   Moamar Sayed-Mouchaweh

Contents

About the Editor

Moamar Sayed-Mouchaweh received his master's degree from the University of Technology of Compiegne-France and PhD degree and the Habilitation to Direct Researches (HDR) in Computer science, Control and Signal processing from the University of Reims-France. Since September 2011, he is working as a Full Professor in the High National Engineering School of Mines-Telecom Lille-Douai in France. He edited and wrote several books for Springer, served as a member of Editorial Boards, International Programme Committee (IPC), conference, workshop and tutorial chair for different international conferences, an invited speaker, and a guest editor of several special issues of international journals targeting the use of advanced artificial intelligence techniques and tools for digital transformation (energy transition and industry 4.0). He served and is serving as an expert for the evaluation of industrial and research projects in the domain of digital transformation. He is leading an interdisciplinary and industry-based research theme around the use of advanced Artificial Intelligence techniques in order to address the challenges of energy transition and Industry 4.0.

Chapter 1
Prologue: Introduction to Explainable Artificial Intelligence

Moamar Sayed-Mouchaweh

1.1 Explainable Machine Learning

Machine learning methods, especially deep neural networks, are becoming increasingly popular in a large variety of applications. These methods learn from observations in order to build a model that is used to generalize prediction (classification or regression) to unknown data. For instance in the healthcare domain, a model is learned in order to decide whether a patient has a cancer or not by treating its microscopic radio images as input. Other example is credit scoring where a model is used to decide whether a candidate may obtain a loan or not.

Machine learning methods generate or learn a highly effective mapping between input and output. Hence, they behave as "black box" entailing a huge difficulty for humans to understand how and why the prediction (output) was made. However in many applications, it is important to explain to users how the decision (prediction) was made by the model and its meaning using understandable terms. Indeed, explainable models allow users to trust them and to better use them thanks to the detailed information (explanation) on how and why they arrived to the provided prediction. Therefore, making machine learning models transparent to human practitioners or users leads to new types of data-driven insights.

Explainable Artificial Intelligence (XAI) [1] aims at producing explainable models that enable human users to understand and appropriately trust the obtained results. The produced explanations allow to reveal how the model functions, why it behaves that way in the past, present and future, why certain actions were taken or must be taken, how certain goals can be achieved, how the system reacts to certain inputs or actions, what are the causes for the occurrence of a certain fault

M. Sayed-Mouchaweh (✉)
Institute Mines-Telecom Lille Douai, Douai, France
e-mail: moamar.sayed-mouchaweh@mines-douai.fr

and how this occurrence can be avoided in the future, etc. The need for explanations is increasingly becoming necessary in multiple application domains, such as smart grids [2], autonomous cars [3], smart factory or industry 4.0 [4, 5], telemedicine and healthcare [6], etc., in particular within the context of digital transformation and cyber-physical systems [7]. For instance in credit scoring or loan approval system, providing an explanation why a loan was refused to applicants allows bias detection. In cybersecurity [8], explanations may help to ensure the security and safety of a system by determining abnormal changes in its output. These changes were generated by hackers in order to fool the model. Indeed, model transparency allows assessing the quality of output predictions and warding off adversaries [9]. However, explanation is required for critical systems where a mistake may have serious consequences such as for self-drive cars.

Making machine learning-based AI explainable faces several challenges. Firstly, the explanations must be adapted to different stakeholders (end users, policy makers, industries, utilities, etc.) with different levels of technical knowledge (managers, engineers, technicians, etc.) in different application domains. Secondly, it is important to develop an evaluation framework and standards in order to measure the effectiveness of the provided explanations at the human and the technical levels. For instance, this evaluation framework must be able to verify that each explanation is consistent across similar predictions (similar observations) over time, is expressive in order to increase the user confidence (trust) in the decisions made, promote impartial and fair decisions, and improve the user task performance. In order to address these challenges, several questions need to be answered:

- How explanations can be integrated into new and already existing AI systems and models?
- How effective user explanation interfaces can be designed?
- How the social (e.g., user requirements, user confidence) and ethical (e.g., bias detection, fairness) aspects of explainability can be guaranteed and evaluated in the designed XAI models?
- How objective benchmarks, standards, and evaluation strategies can be developed in order to measure the effectiveness of XAI models for different applications, uses, and stakeholders of different expertise levels?

Explanations can be generated either to explain the model output (prediction) or its parameters. For instance for the cancer detection model, explanations can be a map of the microscopic image's pixels that contribute the most to the provided decision or prediction. It can also be the features or activation functions' parameters that contributed or related the most to the provided prediction.

Explanations can be generated either during the system design or during operation in order to ensure its quality and reliability. Explainable systems can be either self-explainable or user-triggered. Self-explainable systems are able to generate explanation whenever something requires to be explained in particular at run-time. User-trigger explanation provides explanation whenever a user requests it. For both cases, a model is required allowing to understand the system's internal dynamics, its goals and requirements, its surrounding environments (contextual knowledge), and the recipient of the explanation.

In general, XAI approaches can be categorized in terms of their scope, methodology, and usage [9]. The scope indicates the focus of the explanation on a local instance [10] or on the model as a whole [11]. The methodology refers to the focus of the used approach on the input data instance [10] or the model parameters [12]. Finally, the usage concerns how the explainability is integrated to the model either to a specific model (intrinsic) or to any model as a post hoc explanation (model-agnostic) [10]. Each of the previous approaches has its advantages and drawbacks. For instance, local explanation is much easier than global explanation since it is easier to provide an explanation about a local instance than for a whole model. Moreover, post hoc agnostic models have the advantage to be dependent of the machine learning algorithm used to train the model and therefore can be applied to any already existing model.

1.2 Beyond State-of-the-Art: Contents of the Book

The following chapters in this book overview, discuss, and propose different explainable AI structures, tools, and methods in order to address the aforementioned challenges in different applications (healthcare, finance, cybersecurity, document summarization).

1.2.1 Chapter 2: Principles of Explainable Artificial Intelligence

This chapter presents a state of the art around the major explainable machine learning (XAI) methods and techniques. It starts firstly by highlighting the motivation of XAI in particular when it is used in automated decision-making process for critical systems such as healthcare or autonomous vehicles. Then, the chapter presents a set of dimensions that can be used to analyze or evaluate the capacity of XAI to explain in understandable terms to a human, so-called interpretability. The chapter divides XAI methods into categories according to their capacity to provide global or local explanations, if the explanation is related to a specific machine learning model or model-agnostic explainers, to the time that a user is allowed to spend on understanding an explanation as well as their capacity to guarantee the safety desiderata such as privacy, secrecy, security, and fairness. Then, the chapter evaluates some well-known explainers according to the aforementioned dimensions and categories, such as Trepan, Rxren, Corels, Lore, and Anchor, explainers based on the use of a saliency map, prototype-based explanations, and counterfactual explanations.

1.2.2 Chapter 3: Science of Data: A New Ladder for Causation

This chapter proposes a theoretic framework to create an explainable AI tool that is capable of reasoning. The chapter starts by discussing how deep neural networks explain the decision they made by finding the causes generating the current events or situations. To this end, the chapter proposes a cognitive architecture (Learning Intelligent Distributed Agents (LIDA)) equipped with probabilistic fuzzy logic and graphical neural networks for reasoning. Probabilistic fuzzy logic treats three types of uncertainty: randomness, probabilistic uncertainty, and fuzziness. Therefore, it can manage the uncertainty of our knowledge (by the use of probabilities) and the vagueness inherent to the world's complexity (by data fuzzification). It is integrated with graphical neural networks for learning since it cannot learn by itself and needs experts to define intervals before applying fuzzification.

1.2.3 Chapter 4: Explainable Artificial Intelligence for Predictive Analytics on Customer Turnover

This chapter presents an interactive explainable artificial intelligence web interface that integrates and enhances the state-of-the-art techniques in order to produce more understandable and practical explanations to nontechnical end users. It is applied for the prediction of a financial institution's customer churn rate. The Churn is the rate of customers who stopped using a service or product in a given time frame. It is used in business sector such as banking, retail, telecommunications, and education. The proposed explainable web interface combines visualization and verbalization. The visual screens display local and global features importance in order to provide users with the relevance of each feature to the decision made on a certain instance or on the global model. The verbalization is used as an alternative explanation other than the one provided by the visual screens. In addition, it is used as a recommendation to what to do in order to prevent a customer from leaving the company. The proposed explanation web interface is evaluated and compared with some well-known similar explanation tools, such as GAMUT, TELEGAM, and XAI Clinical Diagnostic Tool, using the following criteria: global explanation, local explanation, interactive (contrastive), search table, target instance capability, and the targeted audience.

1.2.4 Chapter 5: An Efficient Explainable Artificial Intelligence Model of Automatically Generated Summaries Evaluation

This chapter handles the problem of the evaluation of automatically generated summaries. The latter is used for facilitating the selection process for a document and index documents more efficiently, in particular when dealing with massive textual datasets. The chapter presents the most important cognitive psychology models for text comprehension such as the Resonance Model, Landscape Model, Langston and Trabasso Model, and the Construction Integration Model. Then, it proposes a cognitive protocol for Automatically Generated Summaries Evaluation (AGSE) based on a cognitive psychology model of reading comprehension. The originality of this protocol is that it takes into consideration the extent to which an abstract is a good abstract by using three criteria: retention, fidelity, and coherence. The retention checks whether the generated output covers all the concepts reported in the source document, the fidelity gives insights into the extent to which the generated summary accurately reflects the author's point of view by focusing on salient concepts conveyed in the source text, and the coherence checks if the automatically generated summary is semantically meaningful. The retention and fidelity scores are modeled using three linguistic variables, "low," "medium," and "high" represented as membership functions. Then the score combining both the retention and fidelity through operators OR and AND is calculated as fuzzy score using three rules. For instance, *If the retention score is low and the fidelity score is also low, then, the R-F score is low.* Finally, text units having the highest *R-F* scores after defuzzification will present candidate sentences of an ideal summary. Three datasets containing articles and extractive summaries about different topics, such as crisis or protest or war in some Middle East countries, are used for the evaluation of the presented protocol.

1.2.5 Chapter 6: On the Transparent Predictive Models for Ecological Momentary Assessment Data

This chapter describes the use of some well-known classification methods (Bayesian Network, Gradient-boosted trees, Naïve Bayes, Multi-Layer Perceptron, Random Forest, and Decision Tree) in order to assess individuals' eating behavior. The goal of this classification is to predict the future eating behavior of an individual regarding eating healthy or unhealthy in order to intervene just in time if the future behavior is eating unhealthy. The chapter explains the different steps used in order to perform this prediction as well as its interpretation. The dataset represents samples collected from 135 overweight participants over 8 weeks. The goal of the interpretation step is to understand the triggers that lead users to make a choice, which is less healthy than the others are. The extracted features are categorical, such as the location of eating, and continuous, e.g., craving. These features' capacity to discriminate a certain class from the others is used for the interpretation. Each individual is represented by a circle with a size, color, and edges. The color indicates the cluster type, while the size represents how much the corresponding individual shares behavior patterns with the other individuals belonging to the same cluster. The edges correspond to the links between the different individuals belonging to the same clusters.

1.2.6   Chapter 7: Mitigating the Class Overlap Problem in Discriminative Localization: COVID-19 and Pneumonia Case Study

This chapter treats the problem of distinguishing COVID-19 from other pneumonia within a single model trained to detect both diseases using computed tomography (CT) scans, lung ultrasonography (LUS) imagery, and chest X-rays. The challenge is that both diseases are very close entailing to have overlapped classes in the feature space. In addition, there are much fewer COVID-19 labels to learn from entailing class imbalance problems. Therefore, this chapter proposes an approach, called Amplified Directed Divergence, that works with ensembles of models to deal with class overlap and class imbalance while ensuring confident predictive assessments. This approach does not require localized labels, since they can be labor-intensive, but rather exploits the Class Activation Maps (CAMs) computed at the final convolutional layer for each class in a Convolutional Neural Network (CNN) model. The goal is to perform the classification and localization of COVID-19 Regions of Interest (RoI) from CT scans, LUS imagery, and chest X-rays. The salient regions in the COVID-19 CAMs can then be unsampled to the size of the original image in order to localize the features most conducive to the classification of the chosen class. In order to mitigate aleatoric uncertainty, related to overlapped classes, a kernel method is used. It accepts two class activation maps from expert models, each trained on specific overlapped classes, and extracts activations that are relevant to one of them, i.e., target class (COVID-19). The result is a new class activation map that better localizes objects of interest in the presence of class overlap. The proposed approach is validated using COVID-19 and Viral Pneumonia imagery. The obtained results show that the proposed approach enables machine learning practitioners and subject matter experts in various domains in order to increase their confidence in predictions provided by models trained on image-level labels when object-level labels are not available.

1.2.7 Chapter 8: A Critical Study on the Importance of Feature Selection for Diagnosing Cyber-Attacks in Water Critical Infrastructures

This chapter proposes the use of feature selection techniques in order to improve the cyber-attack detection and classification system. This improvement is achieved thanks to the elimination of irrelevant and redundant features from the original data, in particular when they are described by high dimensional and low-quality feature space. In addition, this elimination allows to reduce learning time and prevent overfitting. The chapter proposes the comparison of 12 feature selection techniques in order to effectively select the optimal set of features for detecting intrusion. Then, the selected features are used by two different supervised classification methods k-Nearest Neighbors (kNN) and Decision Trees (DT) in order to perform the intrusion (cyber-attack) classification. The proposed intrusion detection system is applied to a water storage system. The latter is a cyber-physical system vulnerable to seven different types of attacks. The chapter compares the performance of the 12 feature selection techniques and the impact of selected features on the kNN and DT intrusion detection and classification accuracy. The feature selection and analysis can be seen as a way to identify the features that contribute the most to detect a cyber-attack. Then, these features can be used in order to explain the nature, type, and behavior of that detected type of attack.

1.2.8 Chapter 9: A Study on the Effect of Dimensionality Reduction on Cyber-Attack Identification in Water Storage Tank SCADA Systems

This chapter proposes the use of dimensionality reduction techniques (Locally Linear Embedding, Isomap, Linear Discriminant Analysis, Multidimensional Scaling, Principal Component Analysis, etc.) in order to improve the cyber-attack detection and classification system. This chapter is complementary to the previous chapter in the sense that feature selection usually works when at least a number of features possesses very useful information, while dimensionality reduction tries to rectify the feature space and obtain an improved distribution. Indeed, dimensionality reduction can be very helpful in the design of intrusion detection systems (IDS). For instance, if a cyber-attack can be detected by monitoring a large number of features, dimensionality reduction can yield a feature space in which only one or a very small number of features are enough to explain a change that indicates a cyber-attack. In contrast, other techniques such as feature selection may not result in the same efficiency, as the features may not have enough information to only select a small number of them to detect a cyber-threat. The chapter discusses the benefit of using dimensionality reduction to provide an explanation of the detected intrusion and confidence of the obtained decision (detected intrusion). Indeed, the feature space may contain hidden characteristics that are dormant to human eye. Dimensionality reduction techniques improve the explainability by capturing the complex structure of the original data, and then transforming it into a low-dimensional space, which facilitates visualization, revealing relationships between samples, understanding and monitoring the dynamics of the system.

References

1. Arrieta, A. B., Díaz-Rodríguez, N., Del Ser, J., Bennetot, A., Tabik, S., Barbado, A., & Herrera, F. (2020). Explainable artificial intelligence (XAI): Concepts, taxonomies, opportunities and challenges toward responsible AI. *Information Fusion, 58*, 82–115.
2. Murray, D., Stankovic, L., & Stankovic, V. (2020, November). Explainable NILM networks. In *Proceedings of the 5th International Workshop on non-intrusive load monitoring* (pp. 64–69).
3. Li, Y., Wang, H., Dang, L. M., Nguyen, T. N., Han, D., Lee, A., & Moon, H. (2020). A deep learning-based hybrid framework for object detection and recognition in autonomous driving. *IEEE Access, 8*, 194228–194239.
4. Lughofer, E., & Sayed-Mouchaweh, M. (Eds.). (2019). *Predictive maintenance in dynamic systems: Advanced methods, decision support tools and real-world applications.* Springer Nature.
5. Christou, I. T., Kefalakis, N., Zalonis, A., & Soldatos, J. (2020, May). Predictive and explainable machine learning for industrial internet of things applications. In *2020 16th International Conference on Distributed Computing in Sensor Systems (DCOSS)* (pp. 213–218). IEEE.
6. Holzinger, A., Malle, B., Kieseberg, P., Roth, P. M., Müller, H., Reihs, R., & Zatloukal, K. (2017). Towards the augmented pathologist: Challenges of explainable-AI in digital pathology. *arXiv preprint arXiv:1712.06657.*
7. Sayed-Mouchaweh, M. (Ed.). (2020). *Artificial intelligence techniques for a scalable energy transition: Advanced methods, digital technologies, decision support tools, and applications.* Springer Nature.
8. Mahbooba, B., Timilsina, M., Sahal, R., & Serrano, M. (2021). Explainable artificial intelligence (XAI) to enhance trust management in intrusion detection systems using decision tree model. *Complexity, 2021.*
9. Das, A., & Rad, P. (2020). Opportunities and challenges in explainable artificial intelligence (XAI): A survey. *arXiv preprint arXiv:2006.11371.*
10. Zeiler, M. D., & Fergus, R. (2014). Visualizing and understanding convolutional networks. In *European conference on computer vision* (pp. 818–833). Springer.
11. Letham, B., Rudin, C., McCormick, T. H., & Madigan, D. (2015). Interpretable classifiers using rules and Bayesian analysis: Building a better stroke prediction model. *Annals of Applied Statistics, 9*(3), 1350–1371.
12. Ribeiro, M. T., Singh, S., & Guestrin, C. (2016). "Why should I trust you?" explaining the predictions of any classifier. In *Proceedings of the 22nd ACM SIGKDD International Conference on knowledge discovery and data mining* (pp. 1135–1144).

Chapter 2
Principles of Explainable Artificial Intelligence

Riccardo Guidotti, Anna Monreale, Dino Pedreschi, and Fosca Giannotti

2.1 Introduction

Artificial Intelligence is nowadays one of the most important scientific and technological areas, with a tremendous socio-economic impact and a pervasive adoption
in every field of the modern society. High-profile applications such as medical
diagnosis, spam filtering, autonomous vehicles, voice assistants, and image recognition are based on Artificial Intelligence (AI) systems. These AI systems reach
their impressive performance mainly through *obscure* machine learning models that
"hide" the logic of their internal decision processes to humans because they are not
humanly understandable. *Black box models* are tools used by AI to accomplish a task
for which either the logic of the decision process is not accessible or it is accessible
but not human-understandable. Examples of machine learning black box models
adopted by AI systems include neural networks, deep neural networks, ensemble
classifiers, SVMs, but also compositions of expert systems, data mining, and hard-
coded proprietary software. The choice of using not interpretable machine learning
models in AI systems is due to their high performance in terms of accuracy [71].
As a consequence, we have witnessed the rise of a black box society [54], where AI
systems adopt obscure decision-making models to carry on their decision processes.

   The missing of interpretability on how black box models make decisions and
fulfill their tasks is a crucial issue for ethics and a limitation to AI adoption in
socially sensitive and safety-critical contexts such as healthcare and law. Also, the
problem is not only for lack of transparency but also for possible biases inherited

R. Guidotti (✉) · A. Monreale · D. Pedreschi
University of Pisa, Pisa, Italy
e-mail: riccardo.guidotti@unipi.it; anna.monreale@unipi.it; dino.pedreschi@unipi.it

F. Giannotti
ISTI-CNR Pisa, Pisa, Italy
e-mail: fosca.giannotti@isti.cnr.it

by the black boxes from prejudices and artifacts hidden in the training data used by the obscure machine learning models. Indeed, machine learning models are built through a learning phase on training data. These training datasets can contain data coming from the digital traces that people produce while performing daily activities such as purchases, movements, posts in social networks, etc., but also from logs and reports generated in business companies and industries. These "Big Data" can inadvertently contain bias, prejudices, or spurious correlations due to human annotation or the way they are collected and cleaned. Thus, obscure biased models may inherit such biases, possibly causing wrong and unfair decisions. As a consequence, the research in eXplainable AI (XAI) has recently caught much attention [1, 7, 32, 49].

The rest of this chapter is organized as follows. First, Sect. 2.2 shows theoretical, ethical, and legal motivations for the need of an explainable AI. Section 2.3 illustrates the dimensions to distinguish XAI approaches. Then, Sect. 2.4 presents the most common types of explanations and provides some details on the state-of-the-art explanators returning them. Finally, Sect. 2.5 concludes this chapter by discussing practical usability of XAI methods, explanations in real-world applications, and the open research questions.

2.2 Motivations for XAI

Why do we need XAI? In the following, we analyze some real cases depicting how and why AI equipped with black box models can be dangerous both for the possibility of discrimination and for the unavailability of justification after incorrect behaviors.

Prejudices and preconceptions on training datasets can be adopted by machine learning classifiers as general rules to be replicated [56]. Automated discrimination is not necessarily due to black box models. In St. George's Hospital Medical School, London, UK, a program for screening job applicants was used during the 1970s and 1980. The program used information from candidates without any reference to ethnicity. However, such a program was found to discriminate against ethnic minorities and women by inferring this information from surnames and place of birth and lowering their chances of being selected for interview [44]. A more recent example is related to Amazon. In 2016, the AI software used by Amazon to determine the areas of the USA to which Amazon would offer free same-day delivery accidentally restricted minority neighborhoods from participating in the program (often when every surrounding neighborhood was allowed).[1] In the same year, the COMPAS score, a predictive model for the "risk of crime recidivism" (proprietary secret of Northpointe), was shown to have a strong ethnic bias from

---

[1]http://www.techinsider.io/how-algorithms-can-be-racist-2016-4.

the journalists of *propublica.org*.[2] The journalists proved that, according to the COMPAS score, a Black who did not re-offend was classified as "high risk" twice as much as Whites who did not re-offend. On the other hand, White repeat offenders were classified as "low risk" twice as much as Black repeat offenders.

These kinds of biases are tied with the training data. For example, in [15], it is proved that the word embeddings [11] trained on Google News articles exhibit female/male gender stereotypes. Indeed, it is shown that for the analogy "Man is to computer programmer as woman is to X," the variable X was replaced by "homemaker" by the trained obscure model. The problem was the literature and texts used to train the model repeating that a woman does the housework. Similarly, in [58], it is shown that a classifier trained to recognize wolves and husky dogs was basing its predictions to distinguish a wolf solely on the presence of snow in the background. This was happening because all the training images with a wolf had snow in the background. These spurious correlations, biases, and implicit rules, hidden in the data, besides discriminating, can also cause wrong and unfair decisions. Unfortunately, in various cases, machine errors could have been avoided if the AI would not have been obscured. In particular, accessing the reasons for the AI decisions is especially crucial in safety-critical AI systems like medicine and self-driving cars, where a possible erroneous outcome could even lead to the death of people. For example, the incident that involved a self-driving Uber car that knocked down and killed a pedestrian in Tempe, Arizona, in 2018.[3] An appropriate XAI method would have helped the company to understand the reasons behind the decision and manage their responsibilities.

Precisely to avoid these cases, the European Parliament turned into law in May 2018 the *General Data Protection Regulation (GDPR)* containing innovative clauses on interpretability for automated decision-making systems. For the first time, the GDPR introduces a *right of explanation* for all individuals to obtain "meaningful explanations of the logic involved" when automated decision-making takes place. Despite conflicting opinions among legal scholars regarding the real scope of these clauses [27, 47, 73], a joint agreement on the need for the implementation of such a principle is crucial, and it is nowadays a big open scientific challenge. Indeed, without a technology able to explain black box models, the right to explanation will remain a "dead letter." How can companies trust their AI products without fully validating and understanding the rationale of their obscure models? And in turn, how can users trust AI services and applications? It would be unthinkable to increase the trust of people and companies in AI without explaining to humans the logic followed by black box models. For these reasons, XAI is at the heart of responsible, open data science across multiple industry sectors and scientific disciplines involving robotics, economics, sociology, and psychology besides computer science.

---

[2]http://www.propublica.org/article/machine-bias-risk-assessments-in-criminal-sentencing.

[3]https://www.nytimes.com/2018/03/19/technology/uber-driverless-fatality.html.

2.3 Dimensions of XAI

The goal of XAI is to *interpret* AI reasoning. To *interpret* means to give or provide the meaning or to explain and present in understandable terms some concepts.[4] Therefore, in AI, *interpretability* is defined as the ability to *explain* or to provide the meaning in understandable terms to a human [7, 21]. These definitions assume that the concepts composing an explanation and expressed in the understandable terms are self-contained and do not need further explanations. An explanation is an "interface" between a human and an AI, and it is at the same time both human-understandable and an accurate proxy of the AI.

We can identify a set of dimensions to analyze the interpretability of AI systems that, in turn, reflect on the existing different types of explanations [32]. Some of these dimensions are related to *functional* requirements of explainable Artificial Intelligence, i.e., requirements that identify the algorithmic adequacy of a particular approach for a specific application, while others to the *operational* requirements, i.e., requirements that take into consideration how users interact with an explainable system and what is the expectation. Some dimensions instead derive from the need of *usability* criteria from a user perspective, while others derive from the need of guarantees against any vulnerability issues. Recently, all these requirements have been analyzed [68] to provide a framework allowing the systematic comparison of explainability methods. In particular, in [68], the authors propose *Explainability Fact Sheets* that enable researchers and practitioners to assess capabilities and limitations of a particular explainable method. As an example, given an explanation method m, we can consider the following functional requirements. *(i)* Even though m is designed to explain regressors, can we use it to explain probabilistic classifiers? *(ii)* Can we employ m on categorical features even though it only works on numerical features? On the other hand, as an operational requirement, can we consider which is the *function of the explanation*? Provide transparency, assess the fairness, etc.

Besides the detailed requirements illustrated in [68], in the literature, it is recognized as a categorization of explanation methods among fundamental pillars [1, 32]: *(i)* black box explanation vs. explanation by design, *(ii)* global vs. local explanations, and *(iii)* model-specific vs. model-agnostic explainers. In the following, we present details of these distinctions and other important features characterizing XAI methods. Figure 2.1 illustrates a summarized ontology of the taxonomy used to classify XAI methods.

Black Box Explanation vs. Explanation by Design We distinguish between black box explanation and explanation by design. In the first case, the idea is to couple an AI with a black box model with an explanation method able to interpret the black box decisions. In the second case, the strategy is to substitute the obscure model with a transparent model in which the decision process is accessible by design. Figure 2.2

---

[4]https://www.merriam-webster.com/.

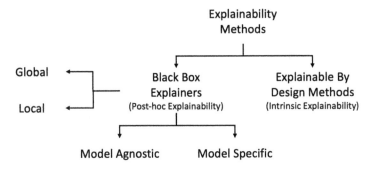

Fig. 2.1 A summarized ontology of the taxonomy of XAI methods

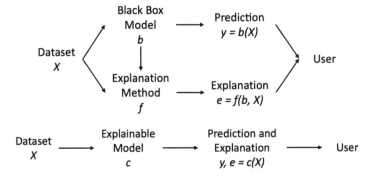

Fig. 2.2 (Top) Black box explanation pipeline. (Bottom) Explanation by design pipeline

depicts this distinction. Starting from a dataset X, the *black box explanation* idea is to maintain the high performance of the obscure model b used by the AI and to use an explanation method f to retrieve an explanation e by reasoning over b and X. This kind of approach is the one more addressed nowadays in the XAI research field [20, 45, 58]. On the other hand, the *explanation by design* consists of directly designing a comprehensible model c over the dataset X, which is interpretable by design and returns an explanation e besides the prediction y. Thus, the idea is to use this transparent model directly into the AI system [61, 62]. In the literature, there are various models recognized to be interpretable. Examples are *decision tree, decision rules*, and *linear models* [24]. These models are considered easily understandable and interpretable for humans. However, nearly all of them sacrifice performance in favor of interpretability. In addition, they cannot be applied effectively on data types such as images or text but only on tabular, relational data, i.e., tables.

Global vs. Local Explanations We distinguish between global and local explanations depending on whether the explanation allows understanding the whole logic of a model used by an AI system or if the explanation refers to a specific case, i.e., only a single decision is interpretable. A *global* explanation consists in providing a way for interpreting any possible decision of a black box *model*. Generally, the black

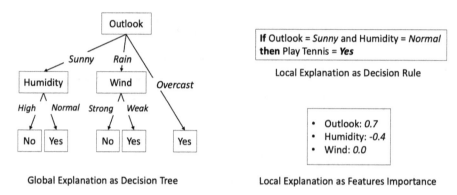

Fig. 2.3 Explanation examples in the form of decision tree, decision rule, and feature importance

box behavior is approximated with a transparent model trained to mimic the obscure model and also to be human-understandable. In other words, the interpretable model approximating the black box provides a global interpretation. Global explanations are quite difficult to achieve and, up to now, can be provided only for AI working on relational data. A *local* explanation consists in retrieving the reasons for the *outcome* returned by a black box model relatively to the decision for a specific instance. In this case, it is not required to explain the whole logic underlying the AI, but only the reason for the prediction on a specific input instance. Hence, an interpretable model is used to approximate the black box behavior only in the "neighborhood" of the instance analyzed, i.e., with respect only to similar instances. The idea is that in such a neighborhood, it is easier to approximate the AI with a simple and understandable model. Regarding Fig. 2.2 (top), a *global* explanation method f uses many instances X over which the explanation is returned. Figure 2.3 (left) illustrates an example of global explanation e obtained by a decision tree structure for a classifier recommending to play tennis or not. The overall decision logic is captured by the tree that says that the classifier recommends playing tennis or not by first looking at the *Outlook* feature. If its value is *Overcast*, then the prediction is "not to play." Otherwise, if its value is *Sunny*, the classifier checks the *Humidity* feature and recommends "not to play" if the Humidity is *High* and "to play" if it is *Normal*. The same reasoning applies to the other branches of the tree. Still with reference to Fig. 2.2 (top), a *local* explanation method f returns an explanation only for a single instance x. Two examples of local explanations are shown in Fig. 2.3 (right). The local rule-based explanation e for a given record x says that the black box b suggested to play tennis because the *Outlook* is *Sunny* and the *Humidity* is *Normal*. On the other hand, the explanation e formed by feature importance says that the black box b suggested playing tennis because the *Outlook* has a large positive contribution, *Humidity* has a consistent negative contribution, and *Wind* has no contribution in the decision.

Interpretable Models for Explaining AI To explain obscure AI systems or to replace the black box components, often interpretable models are learned. The most largely adopted interpretable models are briefly described in the following. A *decision tree* exploits a graph-structure like a tree and composed of internal nodes representing tests on features or attributes (e.g., whether a variable has a value lower than, equal to, or greater than a threshold) and leaf nodes representing a decision. Each branch represents a possible outcome [57]. The paths from the root to the leaves represent the classification rules. The most common rules are *if-then rules*, where the "if" clause is a combination of conditions on the input variables. If the clause is verified, the "then" part reveals the AI action. For a *list of rules*, given an ordered set of rules, the AI returns as the decision the consequent of the first rule that is verified [76]. Finally, *linear models* allow visualizing the *feature importance*: both the sign and the magnitude of the contribution of the attributes for a given prediction [58]. If the sign of an attribute value is positive, then it contributes by increasing the model's output, and otherwise, it decreases it. Higher magnitudes of attribute values indicate a higher influence on the prediction of the model. Examples of such explanations are illustrated in Fig. 2.3.

User's Desiderata Since interpretable models are required to retrieve explanations, some desiderata should be taken into account when adopting them [24]. *Interpretability* consists of evaluating to what extent a given explanation is human-understandable. An approach often used for measuring the interpretability is the *complexity* of the interpretable surrogate model. The complexity is generally estimated with a rough approximation related to the *size* of the interpretable model. For example, the complexity of a rule can be measured with the number of clauses in the condition; for linear models, it is possible to count the number of non-zero weights, while for decision trees the depth of the tree.

Fidelity consists in evaluating to which extent the interpretable model is able to accurately *imitate*, either globally or locally, the decision of the AI. The fidelity can be practically measured in terms of accuracy score, F1-score, etc. [71] with respect to the decisions taken by the black box model. The fidelity has the goal to determine the soundness and completeness of explanations.

Another important property for the user's point view is the *usability*: an interactive explanation can be more useful than a textual and static explanation. However, machine learning models should also have other ordinary important requirements such as *reliability* [42], *robustness* [34], *causality* [28], *scalability*, and *generality* [55]. Reliability and robustness request that an explanation method should have the ability to maintain certain levels of performance independently from small variations of the parameters or of the input. Causality assumes that controlled changes in the input affect the black box behavior in an expected way, known by the explainer. Generality requires that explanation models are portable to different data (with similar nature) without special constraints or restrictions. Finally, since most of the AI systems need "Big Data," it is opportune to have explainers able to scale to large input data.

Moreover, a fundamental aspect is that every explanation should be *personalized* coherently with the user *background*. Different background knowledge and diverse experiences in various tasks are tied to different notions and requirements for the usage of explanations. Domain experts can be able to understand complex explanations, while common users require simple and effective clarifications. Indeed, the meaningfulness and usefulness of an explanation depend on the stakeholder [12]. Taking as an example the aforementioned COMPAS case, a specific explanation for a score may make sense to a judge who wants to understand and double-check the suggestion of the AI support system and possibly discover that it biased against Blacks. On the other hand, the same explanation is not useful to a prisoner who cannot change the reality of being Black, while he can find the suggestion meaningful that when he will be older then he would lower his risk down. Moreover, besides these features strictly related to XAI, an interpretable model should satisfy other important general desiderata. For instance, having a high *accuracy* that consists in evaluating to what extent the model accurately takes decisions for unseen instances.

Model-Specific vs. Model-Agnostic Explainers We distinguish between model-specific and model-agnostic explanation methods depending on whether the technique adopted to retrieve the explanation acts on a particular model adopted by an AI system or can be used on any type of AI. The most used approach to explain AI black boxes is known as *reverse engineering*. The name comes from the fact that the explanation is retrieved by observing what happens to the output, i.e., the AI decision, when changing the input in a controlled way. An explanation method is *model-specific* or not generalizable [48], if it can be used to interpret only particular types of black box models. For example, if an explanation approach is designed to interpret a random forest [71] and internally use a concept of distance between trees, then such an approach cannot be used to explain the predictions of a neural network. On the other hand, an explanation method is *model-agnostic*, or generalizable, when it can be used independently from the black box model being explained. In other words, the AI's internal characteristics are not exploited to build the interpretable model approximating the black box behavior.

Time Limitations The time that the user is allowed to spend on understanding an explanation or is available to do it is a crucial aspect. Obviously, the time availability of a user is strictly related to the scenario where the predictive model has to be used. In some contexts where the user needs to quickly take the decision, e.g., a surgery or an imminent disaster, it is preferable to have an explanation simple and effective. While in contexts where the decision time is not a constraint, such as during a procedure to release a loan, one might prefer a more complex and exhaustive explanation.

Safety Desiderata Explainability methods providing interpretable understanding may reveal partial information about the training data, the internal mechanisms of the models, or their parameters and prediction boundaries [14, 65]. Thus, desiderata such as *privacy* [52], *secrecy, security,* and *fairness* [56] should be considered to avoid skepticism and increase trust. *Fairness* and *privacy* are fundamental desiderata to guarantee the protection of groups against (direct or indirect) discrimination [60] that the interpretable model does not reveal sensitive information about people [3].

2.4   Explanations and Explanators

Increasing research on XAI is bringing to light a wide list of explanations and explanation methods for "opening" black box models. The explanations returned depend on various factors such as the type of task they are needed for and on which type of data the AI system acts, who is the final user of the explanation, if they allow to explain the whole behavior of the AI system (global explanations) or reveal the reasons for the decision only for a specific instance (local explanations). In this section, we review the most used types of explanations and show how some state-of-the-art explanation methods are able to return them. The interest reader can refer to [1, 32] for a complete review of the literature in XAI.

2.4.1   Single Tree Approximation

One of the first approaches introduced to explain neural networks is TREPAN [20]. TREPAN is a global explanation method that is able to model the whole logic of a neural network working on tabular with a single decision tree. The decision tree returned by TREPAN as explanation is a *global transparent surrogate*. Indeed, every path from the root of the tree to a leaf shows the reasons for the final decision reported in the leaf. An example of a decision tree returned by TREPAN is illustrated in Fig. 2.4. This global explanation reveals that the black box first focuses on the value of the feature *rest ECG* and depending on its degree (abnormal, normal, hypertrophy) takes different decisions depending on additional factors such as sex or cholesterol. In particular, TREPAN queries the neural network to induce the decision

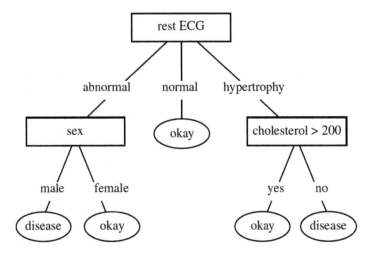

Fig. 2.4 Example of global tree-based explanation returned by TREPAN

tree by maximizing the gain ratio [71] on the data with respect to the predictions of the neural network. A weakness of common decision trees like ID3 or C4.5 [57] is that the amount of data to find the splits near to the leaves is much lower than those used at the beginning. Thus, in order to retrieve how a neural network works in detail, TREPAN adopts a synthetic generation of data that respect the path of each node before performing the splitting such that the same amount of data is used for every split. In addition, it allows flexibility by using *"m-of-n"* rules where only *m* conditions out of *n* are required to be satisfied to classify a record. Therefore, TREPAN maximizes the fidelity of the single tree explanation with respect to the black box decision. We highlight that even though TREPAN is proposed to explain neural networks, in reality it is model-agnostic because it does not exploit any internal characteristic of neural networks to retrieve the explanation tree. Thus, it can be theoretically employed to explain every type of classifier, i.e., it is model-agnostic.

In [16] is presented an extension of TREPAN that aims to keep the tree explanation simple and compact by introducing four splitting approaches aimed at finding the most relevant features during the tree construction. In [36], genetic programming is used to evolve a single decision tree that approximates the behavior of a neural network ensemble by considering additional genetic features obtained as combinations of the original data and the novel data annotated by the black box models. Both methods described in [16, 36] return explanations in the form of a global decision tree. The readers interested can refer to the papers for more details.

2.4.2 Rules List and Rules Set

A decision rule is generally formed by a set of conditions and by a consequent, e.g., *if conditions, then consequent*. Given a record, a decision rule assigns to the record the outcome specified in the consequent if the conditions are verified [2]. The most common rules are *if-then rules* that take into consideration rules with clauses in conjunction. On the other hand, for *m-of-n* rules given a set of *n* conditions, if *m* of them are verified, then the consequence of the rule is applied [51]. When a set of rules is used, then there are different strategies to select the outcome. For a *list of rules*, the order of the list is considered and the model returns the outcome of the first rule that verifies the conditions [76]. For instance, *falling rule lists* are if-then rules ordered with respect to the probability of a specific outcome [75]. On the other hand, *decision sets* are unordered lists of rules. Basically each rule is an independent classifier that can assign its outcome without regard for any other rule [39]. Voting strategies are used to select the final outcome.

List of rules and set of rules are adopted as explanation both from explanation methods and from transparent classifiers. In both cases, the reference context is tabular data. In [8], the explanation method RXREN unveils with rules list the logic behind a trained neural network. First, RXREN prunes the insignificant input neurons and identifies the data range necessary to classify the given test instance with a

if $((data(I_1) \geq L_{13} \wedge data(I_1) \leq U_{13}) \wedge (data(I_2) \geq L_{23} \wedge data(I_2) \leq U_{23}) \wedge$
$(data(I_3) \geq L_{33} \wedge data(I_3) \leq U_{33}))$ **then class** $= C_3$
else
if $((data(I_1) \geq L_{11} \wedge data(I_1) \leq U_{11}) \wedge (data(I_3) \geq L_{31} \wedge data(I_3) \leq U_{31}))$
then class $= C_1$
else
class $= C_2$

Fig. 2.5 Example of the list of rules explanation returned by RXREN

> **if** $(age = 18 - 20)$ **and** $(sex = male)$ **then predict** yes
> **else if** $(age = 21 - 23)$ **and** $(priors = 2 - 3)$ **then predict** yes
> **else if** $(priors > 3)$ **then predict** yes
> **else predict** no

Fig. 2.6 Example of the list of rules explanation returned by CORELS

specific class. Second, RXREN generates the classification rules for each class label exploiting the data ranges previously identified and improves the initial list of rules by a process that prunes and updates the rules. Figure 2.5 shows an example of rules returned by RXREN. A survey on techniques extracting rules from neural networks is [4]. All the approaches in [4], including RXREN, are model-specific explainers.

As previously mentioned, an alternative line of research to black box explanation is the design of transparent models for the AI systems. The CORELS method [5] is a technique for building rule lists for discretized tabular datasets. CORELS provides an optimal and certifiable solution in terms of rule lists. An example of rules list returned by CORELS is reported in Fig. 2.6. The rules are read one after the other, and the AI would take the decision of the first rule for which the conditions are verified. Decision sets are built by the method presented in [39]. The if-then rules extracted for each set are accurate, non-overlapping, and independent. Since each rule is independently applicable, decision sets are simple, succinct, and easily to be interpreted. A decision set is extracted by jointly maximizing the interpretability and predictive accuracy by means of a two-step approach using frequent itemset mining and a learning method to select the rules. The method proposed in [63] merges local explanation rules into a unique global weighted rule list by using a scoring system.

2.4.3 Partial Dependency

Another global XAI method for inspecting the behavior of black box models is the partial dependence plot (PDP). In [32], the black box inspection problem is defined as providing a representation for understanding why the black box returns certain predictions more likely than others with particular inputs. The PDP visually shows the relationship between the AI decision and the input variables in a reduced feature space clarifying whether the relationship is linear, monotonic, or more complex.

In particular, a PDP shows the marginal effect of a feature on the AI decision [25]. Shortly, a feature is selected and it is varied in its domain. Then, instances are built with values of the selected feature and values from the other features of a given training data. The PDP for a value of the selected feature is the mean probability of classification over the training data or the average regression value. An assumption of the PDP is that the selected feature is not correlated with the other features. Generally, PDP approaches are model-agnostic and used on tabular datasets. In [38], the PROSPECTOR method implementing a PDP is proposed to observe how the output of a black box varies by varying the input changing one variable at a time with an effective way to understand which are the most important features. Figure 2.7 shows the PROSPECTOR PDP for the feature *age* and a black box that predicts the risk of churn. In this example, the PROSPECTOR PDP shows the marginal effect (black line) of the feature *Age* on the predicted outcome *Risk* of a black box classifier. In particular, in this case, the higher is the *Age*, the higher is the probability of *Risk of Churn*. We highlight that for *Age* greater than fifty five this probability markedly increases.

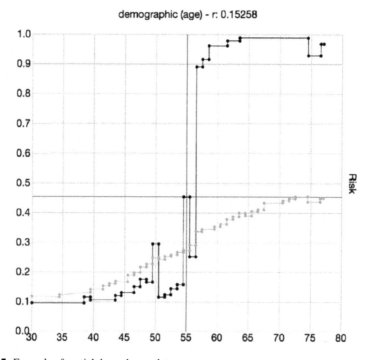

Fig. 2.7 Example of partial dependence plot

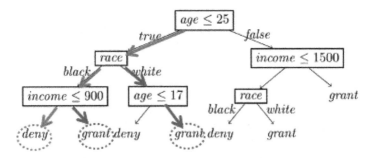

Fig. 2.8 Example of local factual and counterfactual explanation returned by LORE

2.4.4 Local Rule-Based Explanation

Despite being useful, global explanations can be inaccurate because interpreting a whole model can be complex. On the other hand, even though the overall decision boundary is difficult to explain, locally, in the surrounding of a specific instance, it can be easier. Therefore, a local explanation rule can reveal the *factual reasons* for the decision taken by the black box of an AI system for a specific instance. The LORE method is able to return a local rule-based explanation. LORE builds a local decision tree in the neighborhood of the instance analyzed [30] generated with a genetic procedure to account for both similarity and differences with the instance analyzed. Then, it extracts from the tree a local rule revealing the reasons for the decision on the specific instance (see the green path in Fig. 2.8). For instance, the explanation of LORE for the denied request of a loan from a customer with "age=22, race=black, and income=800" to a bank that uses an AI could be the factual rule *if age ≤ 25 and race = black and income ≤900 then deny*. ANCHOR [59] is another XAI approach for locally explaining black box models with decision rules called anchors. An *anchor* is a set of features with thresholds indicating the values that are fundamental for obtaining a certain decision of the AI. An ANCHOR efficiently explores the black box behavior by generating random instances exploiting a multi-armed bandit formulation.

2.4.5 Feature Importance

A widely adopted form of local explanation, especially for tabular data, consists of *feature importance*. Local explanations can also be returned in the form of feature importance that considers both the sign and the magnitude of the contribution of the features for a given AI decision. If the value of a feature is positive, then it contributes by increasing the model's output; if the sign is negative, then the feature decreases the output of the model. If a feature has a higher contribution than another, then it means that it has a stronger influence on the prediction of the black box outcome. The feature importance summarizes the decision of the black box model providing the opportunity of quantifying the changes of the black box decision for each test record. Thus, it is possible to identify the features leading to a certain outcome for a specific instance and how much they contributed to the decision.

The LIME model-agnostic local explanation method [58] randomly generates synthetic instances around the record analyzed and then returns the feature importance as the coefficient of a linear regression model adopted as a local surrogate. The synthetic instances are weighted according to their proximity to the instance of interest. The Lasso model is trained to approximate the probability of the decision of the black box in the synthetic neighborhood of the instance analyzed. Figure 2.9 shows the feature importance returned by LIME (central part of the figure) toward the two classes. In this example, the feature *odor=foul* has a positive contribution of 0.26 in the prediction of a mushroom as *poisonous, stack-surface-above-ring=silky* has a positive contribution of 0.11, *spore-print-color=chocolate* has a positive contribution of 0.08, *stack-surface-below-ring=silky* has a positive contribution of 0.06, while *gill-size=broad* has a negative contribution of 0.13. Another widely adopted model-agnostic local explanation method is SHAP [45]. SHAP connects game theory with local explanations exploiting the *Shapely values* of a conditional expectation function of the black box to explain the AI. Shapley values are introduced in [64] with a method for assigning "payouts" to "players" depending on their contribution to the "total payout." Players cooperate in a coalition and receive a certain "profit" from this cooperation. The connection with explainability is as follows. The "game" is the decision of the black box for a specific instance. The "profit" is the actual value of the decision for this instance minus the average values for all instances. The "players" are the feature values of the instance that leads toward a certain value, i.e., collaborate to receive the profit. Thus, a Shapley value is the *average marginal contribution* of a feature value across all possible coalitions, i.e., combinations [50]. Therefore, SHAP returns the local unique additive feature importance for each specific record. The higher is a Shapely value, and the higher is the contribution of the feature. Figure 2.10 illustrates an example of SHAP explanation, where the feature importance is expressed in the form of a *force plot*. This explanation shows for each feature the level of the contribution in pushing the black box prediction from the base value (the average model output over the

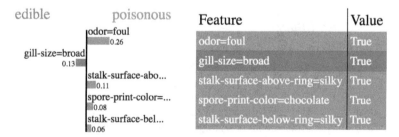

Fig. 2.9 Example of explanation based on feature importance by LIME

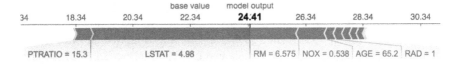

Fig. 2.10 Example of explanation based on feature importance by SHAP

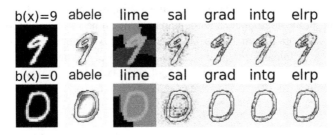

Fig. 2.11 Example of saliency maps returned by different explanation methods. The first column contains the image analyzed and the label assigned by the black box model *b* of the AI system

dataset, which is 24.41 in this example) to the model output. The features pushing the prediction higher are shown in red; those pushing the prediction lower are shown in blue. Under appropriate settings, LIME and SHAP can also be used to explain the decisions of AI working on textual data and images.

2.4.6 Saliency Maps

The most used type of explanation for explaining AI systems working on images consists of *saliency maps*. A saliency map is an image where each pixel's color represents a value modeling the importance of that pixel for the prediction, i.e., they show the positive (or negative) contribution of each pixel to the black box outcome. Thus, saliency maps are returned by local explanation methods. In the literature, there exist different explanation methods locally explaining deep neural networks for image classification. The two most used model-specific techniques are *gradient attribution methods* like SAL [67], GRAD [66], INTG [69], ELRP [9], and *perturbation-based attribution methods* [23, 77]. Without entering into the details, these XAI approaches aim to assign an importance score to each pixel such that the probability of the deep neural network of labeling the image with a different outcome is minimized, if only the most important pixels are considered. Indeed, the areas retrieved by these methods are also called *attention areas*.

The aforementioned XAI methods are specifically designed for specific DNN models, i.e., they are model-specific. However, under appropriate image transformations that exploit the concept of "superpixels" [58], the model-agnostic explanation methods such as LIME, ANCHOR, and LORE can also be employed to explain

AI working on images for any type of black box model. The attention areas of explanations returned by these methods are tied to the technique used for segmenting the image to explain and to a neighborhood consisting of unrealistic synthetic images with "suppressed" superpixels [29]. On the other hand, the local model-agnostic explanation method ABELE [31] exploits a generative model, i.e., an adversarial autoencoder[46], to produce a realistic synthetic neighborhood that allows retrieving more understandable saliency maps. Indeed, ABELE's saliency maps highlight the contiguous attention areas that can be varied while maintaining the same classification from the black box used by the AI system. Figure 2.11 reports a comparison of saliency maps for the classification of the handwritten digits "9" and "0" for the explanation methods ABELE [31], LIME [58], SAL [67], GRAD [66], INTG [69], and ELRP [9].

2.4.7 Prototype-Based Explanations

Prototype-based explanation methods return as explanation a selection of particular instances of the dataset for locally explaining the behavior of the AI system [50]. Prototypes (or exemplars) make clear to the user the reasons for the AI system's decision. In other words, prototypes are used as a foundation of representation of a category, or a concept [26]. A concept is represented through a specific instance. Prototypes help humans in constructing mental models of the black box model and of the training data used. Prototype-based explainers are generally local methods that can be used independently for tabular data, images, and text. Obviously, prototype-based explanations only make sense if an instance of the data is humanly understandable and makes sense as an explanation. Hence, these methods are particularly useful for images, short texts, and tabular data with few features.

In [13], prototypes are selected as a minimal subset of samples from the training data that can serve as a condensed view of a dataset. Naive approaches for selecting prototypes use the closest neighbors from the training data with respect to a predefined distance function, or the closest centroids returned by a clustering algorithm [71]. In [43], we designed a sophisticated model-specific explanation method that directly encapsulates in a deep neural network architecture an autoencoder and a special prototype layer, where each unit of that layer stores a weight vector that resembles an encoded training input. The autoencoder permits to make comparisons within the latent space and to visualize the learned prototypes such that besides accuracy and reconstruction error, the optimization has a term that ensures that every encoded input is close to at least one prototype. Thus, the distances in the prototype layer are used for the classification such that each prediction comes with an explanation corresponding to the closest prototype. In [18], prototypical parts of images are extracted by a PROTOPNET network, such that each classification is driven by prototypical aspects of a class.

Fig. 2.12 Example of exemplars (left) and counter-exemplars (right) returned by ABELE. On top of each (counter-)exemplar is reported the label assigned by the black box model b of the AI system

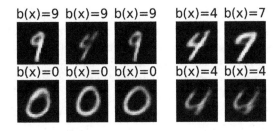

Although prototypes are representative of all the data, sometimes they are not enough to provide evidence for the classification without instances that are not well represented by the set of prototypes [50]. These instances are named *criticisms* and help to explain what is *not* captured by prototypes. In order to aid human interpretability, in [37], prototypes and criticisms are selected by means of the *Maximum Mean Discrepancy* (MMD): instances in highly dense areas are good prototypes, and instances that are in regions that are not well explained by the prototypes are selected as criticisms. Finally, the ABELE method [31] enforces the saliency map explanation with a set of exemplar and counter-exemplar images, i.e., images similar to the one under investigation classified for which the same decision is taken by the AI, and images similar to the one explained for which the black box of the AI returns a different decision. The particularity of ABELE is that it does not select exemplars and counter-exemplars from the training set, but it generates them synthetically exploiting an adversarial autoencoder used during the explanation process [40]. An example of exemplars (left) and counter-exemplars (right) is shown in Fig. 2.12.

2.4.8 Counterfactual Explanations

A *counterfactual* explanation shows what should have been different to change the decision of an AI system. Counterfactual explanations are becoming an essential component in XAI methods and applications [6] because they help people in reasoning on the cause–effect relations between analyzed instances and AI decision [17]. Indeed, while direct explanations such as feature importance, decision rules, and saliency maps are important for understanding the reasons for a certain decision, a counterfactual explanation reveals what should be different in a given instance to obtain an alternative decision [73]. Thinking in terms of "counterfactuals" requires the ability to figure a hypothetical causal situation that contradicts the observed one [50]. Thus, the "cause" of the situation under investigation are the features describing the situation and that "caused" a certain decision, while the "event" models the decision.

The most used types of counterfactual explanations are indeed prototype-based counterfactuals. In [74], counterfactual explanations are provided by an explanation method that solves an optimization problem that, given an instance under analysis, a training dataset, and a black box function, returns an instance similar to the input one and with minimum changes, i.e., minimum distance, but that reverts the black box outcome. The counterfactual explanation describes the smallest change that can be made in that particular case to obtain a different decision from the AI. In [72] is proposed the generation of diverse counterfactuals using mixed integer programming for linear models. As previously mentioned, ABELE [31] also returns synthetic *counter-exemplar images* that highlight the similarities and differences between images leading to the same decision and images leading to other decisions.

Another modeling for counterfactual explanations consists of the logical form that describes a causal situation like: "If X had not occurred, Y would not have occurred" [50]. The local model-agnostic LORE explanation method [30], besides a factual explanation rule, also provides a set of *counterfactual rules* that illustrate the logic used by the AI to obtain a different decision with minimum changes. For example, in Fig. 2.8, the set of counterfactual rules is highlighted in purple and shows that *if income >900 then grant*, or *if race = white then grant*, clarifying which particular changes would revert the decision. In [41] is proposed a local neighborhood generation method based on a Growing Spheres algorithm that can be used for both finding counterfactual instances and acting as a base for extracting counterfactual rules.

2.5   Conclusions

This chapter has discussed the problem of interpretability of AI-based decision systems that typically are opaque and hard to understand by humans. In particular, we have analyzed the different dimensions of the problem and the different types of explanations offered by methods proposed by the scientific community. The opportunity to explain complex AI-based systems is fundamental for the diffusion and adoption of those systems in critical domains. One of the most critical ones is the healthcare field where the question of interpretability is far from just intellectual curiosity. The point is that these systems should be used as a support for physicians who have important responsibilities when taking decisions that have a direct impact on health status of humans. For instance, a XAI system, providing details in the form of logical rules or feature importance, could be extremely useful to medical experts who have to monitor and predict the disease evolution of a patient (diabetes detection [70], Alzheimer progression [53], etc.) while understanding the reason for a specific evolution, progression, and complication. Exactly for studying progression and complication, prototype-based explanations and counterfactual explanations can play a crucial role. On the other hand, exemplars and counter-exemplars could be fundamental for identifying brain tumor by comparing with images from magnetic resonance scans [10] and for highlighting through saliency maps the areas of the brain responsible for the decision of the AI system. These are the only examples because there are many other different cases where the knowledge of the medical staff can be augmented by the knowledge acquired by the machine learning system able to elaborate and analyze myriad of the available information.

Another important field where explainability is applicable is in the context of recommendation systems for getting explainable e-commerce recommendations, explainable social recommendations, and explainable multimedia recommendations. In this context, the goal is to inscribe transparency in the systems but also to provide explanations to final users or system designers who are naturally involved in the loop. In e-commerce, the goal is to explain the ranking of specific recommendations of products [19, 35]. Explainable recommendations also apply to social networks for friend recommendations, recommendation of music, news, travels, tags in images, etc. A useful explanation for recommendation systems could be based on feature importance revealing which are the items contributing positively or negatively to the recommendation. Explainability in social environments is important to increase the users' trustworthiness in the recommendations that is fundamental for the social network sustainability. For instance, in [33], a classifier for predicting the risk of car crash of a driver is equipped with the SHAP explainer that reveals the importance of the features recognizing the risk of collision. Understanding the reasons of recommendations is crucial because it makes the user aware about the technology he/she is using and also about his/her online behavior that enabled the specific recommendation.

Unveiling and interpreting the lending decisions made by an AI-based system is fundamental for the legal point of view and for increasing the social acceptance of these systems. Indeed, these systems based on machine learning models pick up *biases* from the training data. This can lead to learn possible discriminatory behavior against protected groups. In these contexts, interpretability can help in the debugging aimed at detecting those biases and to understand how to have a model able to minimize loan defaults, but also to avoid the discrimination due to certain demographics biases [22]. As a consequence, explainable AI in this setting has a double goal: providing clarification to end user about the reason of the final decisions and providing automated feedback to constantly improve the AI system to eliminate possible ethical issues.

The application domains just discussed are only some of the possible applications of explainable AI. With the AI research advancements, the need of explainability will tend to increase more and more because the complexity of the models could jeopardize their usability. Clearly, the research on explainable AI requires still some effort especially in terms of *personalized and interactive explanations*, i.e., in the study of methods able to provide explanations adaptable to the user background and enabling the human interaction creating the beneficial loop human-machine that could lead the machine to learn from humans and humans from machine.

Acknowledgments This work is partially supported by the European Community H2020 programme under the funding schemes: INFRAIA-01-2018-2019 Res. Infr. G.A. 871042 *SoBigData++* (sobigdata.eu), G.A. 952026 *Humane AI Net* (humane-ai.eu), G.A. 825619 *AI4EU* (ai4eu.eu), G.A. 952215 *TAILOR* (tailor.eu), and the ERC-2018-ADG G.A. 834756 "*XAI*: Science and technology for the eXplanation of AI decision making" (xai.eu).

References

1. Adadi, A., & Berrada, M. (2018). Peeking inside the black-box: A survey on explainable artificial intelligence (XAI). *IEEE Access, 6*, 52138–52160.
2. Agrawal, R., Srikant, R., et al. (1994). Fast algorithms for mining association rules. In *Proc. 20th Int. Conf. Very Large Data Bases, VLDB* (Vol. 1215, pp. 487–499).
3. Aldeen, Y. A. A. S., Salleh, M., & Razzaque, M. A. (2015). A comprehensive review on privacy preserving data mining. *SpringerPlus, 4*(1), 694.
4. Andrews, R., Diederich, J., & Tickle, A. B. (1995). Survey and critique of techniques for extracting rules from trained artificial neural networks. *Knowledge-Based Systems, 8*(6), 373–389.
5. Angelino, E., Larus-Stone, N., Alabi, D., Seltzer, M., & Rudin, C. (2017). Learning certifiably optimal rule lists. In *Proceedings of the 23rd ACM SIGKDD International Conference on Knowledge Discovery and Data Mining* (pp. 35–44). ACM.
6. Apicella, A., Isgrò, F., Prevete, R., & Tamburrini, G. (2019). Contrastive explanations to classification systems using sparse dictionaries. In *International Conference on Image Analysis and Processing* (pp. 207–218). Springer.
7. Arrieta, A. B., Díaz-Rodríguez, N., Del Ser, J., Bennetot, A., Tabik, S., Barbado, A., García, S., Gil-López, S., Molina, D., Benjamins, R., et al. (2020). Explainable artificial intelligence (XAI): Concepts, taxonomies, opportunities and challenges toward responsible AI. *Information Fusion, 58*, 82–115.
8. Augasta, M. G., & Kathirvalavakumar, T. (2012). Reverse engineering the neural networks for rule extraction in classification problems. *Neural Processing Letters, 35*(2), 131–150.
9. Bach, S., Binder, A., Montavon, G., Klauschen, F., Müller, K.-R., & Samek, W. (2015). On pixel-wise explanations for non-linear classifier decisions by layer-wise relevance propagation. *PLoS One, 10*(7), e0130140.
10. Bakas, S., et al. (2018). Identifying the best machine learning algorithms for brain tumor segmentation, progression assessment, and overall survival prediction in the brats challenge.
11. Bengio, Y., Ducharme, R., Vincent, P., & Jauvin, C. (2003). A neural probabilistic language model. *Journal of Machine Learning Research, 3*, 1137–1155.
12. Bhatt, U., Xiang, A., Sharma, S., Weller, A., Taly, A., Jia, Y., Ghosh, J., Puri, R., Moura, J. M., & Eckersley, P. (2020). Explainable machine learning in deployment. In *Proceedings of the 2020 Conference on Fairness, Accountability, and Transparency* (pp. 648–657).
13. Bien, J., & Tibshirani, R. (2011). Prototype selection for interpretable classification. *The Annals of Applied Statistics, 5*(4), 2403–2424.
14. Blanco-Justicia, A., Domingo-Ferrer, J., Martínez, S., & Sánchez, D. (2020). Machine learning explainability via microaggregation and shallow decision trees. *Knowledge-Based Systems, 194*, 105532.
15. Bolukbasi, T., Chang, K.-W., Zou, J. Y., Saligrama, V., & Kalai, A. T. (2016). Man is to computer programmer as woman is to homemaker? Debiasing word embeddings. In *Advances in Neural Information Processing Systems* (pp. 4349–4357).
16. Boz, O. (2002). Extracting decision trees from trained neural networks. In *Proceedings of the Eighth ACM SIGKDD International Conference on Knowledge Discovery and Data Mining* (pp. 456–461).
17. Byrne, R. M. (2019). Counterfactuals in explainable artificial intelligence (XAI): Evidence from human reasoning. In *IJCAI* (pp. 6276–6282).
18. Chen, C., Li, O., Tao, D., Barnett, A., Rudin, C., & Su, J. K. (2019). This looks like that: Deep learning for interpretable image recognition. In *Advances in Neural Information Processing Systems* (pp. 8930–8941).
19. Chen, X., Chen, H., Xu, H., Zhang, Y., Cao, Y., Qin, Z., & Zha, H. (2019). Personalized fashion recommendation with visual explanations based on multimodal attention network: Towards visually explainable recommendation. In B. Piwowarski, M. Chevalier, É. Gaussier, Y. Maarek, J. Nie & F. Scholer (Eds.), *Proceedings of the 42nd International ACM SIGIR Conference on Research and Development in Information Retrieval, SIGIR 2019, Paris, France, July 21–25, 2019* (pp. 765–774). ACM.

20. Craven, M., & Shavlik, J. W. (1996). Extracting tree-structured representations of trained networks. In *Advances in Neural Information Processing Systems* (pp. 24–30).

21. Doshi-Velez, F., & Kim, B. (2017). Towards a rigorous science of interpretable machine learning. arXiv preprint arXiv:1702.08608.

22. Fahner, G. (2018). Developing transparent credit risk scorecards more effectively: An explainable artificial intelligence approach. *Data Anal, 2018*, 17.

23. Fong, R. C., & Vedaldi, A. (2017). Interpretable explanations of black boxes by meaningful perturbation. In *Proceedings of the IEEE International Conference on Computer Vision* (pp. 3429–3437).

24. Freitas, A. A. (2014). Comprehensible classification models: A position paper. *ACM SIGKDD Explorations Newsletter, 15*(1), 1–10.

25. Friedman, J. H. (2001). Greedy function approximation: A gradient boosting machine. *Annals of Statistics, 29*(5), 1189–1232.

26. Frixione, M., & Lieto, A. (2012). Prototypes vs exemplars in concept representation. In *KEOD* (pp. 226–232).

27. Goodman, B., & Flaxman, S. (2016). EU regulations on algorithmic decision-making and a "right to explanation". In *ICML Workshop on Human Interpretability in Machine Learning (WHI 2016), New York, NY*. http://arxiv.org/abs/1606.08813v1

28. Goyal, Y., Feder, A., Shalit, U., & Kim, B. (2019). Explaining classifiers with causal concept effect (CACE). arXiv preprint arXiv:1907.07165.

29. Guidotti, R., Monreale, A., & Cariaggi, L. (2019). Investigating neighborhood generation methods for explanations of obscure image classifiers. In *Pacific-Asia Conference on Knowledge Discovery and Data Mining* (pp. 55–68). Springer.

30. Guidotti, R., Monreale, A., Giannotti, F., Pedreschi, D., Ruggieri, S., & Turini, F. (2019). Factual and counterfactual explanations for black box decision making. *IEEE Intelligent Systems, 34*(6), 14–23.

31. Guidotti, R., Monreale, A., Matwin, S., & Pedreschi, D. (2019). Black box explanation by learning image exemplars in the latent feature space. In *Joint European Conference on Machine Learning and Knowledge Discovery in Databases* (pp. 189–205). Springer.

32. Guidotti, R., Monreale, A., Ruggieri, S., Turini, F., Giannotti, F., & Pedreschi, D. (2018). A survey of methods for explaining black box models. *ACM Computing Surveys (CSUR), 51*(5), 1–42.

33. Guidotti, R., & Nanni, M. (2020). Crash prediction and risk assessment with individual mobility networks. In *2020 21st IEEE International Conference on Mobile Data Management (MDM)* (pp. 89–98). IEEE.

34. Guidotti, R., & Ruggieri, S. (2019). On the stability of interpretable models. In *2019 International Joint Conference on Neural Networks (IJCNN)* (pp. 1–8). IEEE.

35. He, X., Chen, T., Yen Kan, M., & Chen, X. (2015). TriRank: Review-aware explainable recommendation by modeling aspects.

36. Johansson, U., & Niklasson, L. (2009). Evolving decision trees using oracle guides. In *2009 IEEE Symposium on Computational Intelligence and Data Mining* (pp. 238–244). IEEE.

37. Kim, B., Khanna, R., & Koyejo, O. O. (2016). Examples are not enough, learn to criticize! criticism for interpretability. In *Advances in Neural Information Processing Systems* (pp. 2280–2288).

38. Krause, J., Perer, A., & Ng, K. (2016). Interacting with predictions: Visual inspection of black-box machine learning models. In *Proceedings of the 2016 CHI Conference on Human Factors in Computing Systems* (pp. 5686–5697).

39. Lakkaraju, H., Bach, S. H., & Leskovec, J. (2016). Interpretable decision sets: A joint framework for description and prediction. In *Proceedings of the 22nd ACM SIGKDD International Conference on Knowledge Discovery and Data Mining* (pp. 1675–1684). ACM.

40. Lampridis, O., Guidotti, R., & Ruggieri, S. (2020). Explaining sentiment classification with synthetic exemplars and counter-exemplars. In *International Conference on Discovery Science* (pp. 357–373). Springer.
41. Laugel, T., Lesot, M.-J., Marsala, C., Renard, X., & Detyniecki, M. (2017). Inverse classification for comparison-based interpretability in machine learning. arXiv preprint arXiv:1712.08443.
42. Laugel, T., Lesot, M.-J., Marsala, C., Renard, X., & Detyniecki, M. (2019). Unjustified classification regions and counterfactual explanations in machine learning. In *Joint European Conference on Machine Learning and Knowledge Discovery in Databases* (pp. 37–54). Springer.
43. Li, O., Liu, H., Chen, C., & Rudin, C. (2018). Deep learning for case-based reasoning through prototypes: A neural network that explains its predictions. In *Thirty-Second AAAI Conference on Artificial Intelligence*.
44. Lowry, S., & Macpherson, G. (1988). A blot on the profession. *British Medical Journal (Clinical Research Ed.), 296*(6623), 657.
45. Lundberg, S. M., & Lee, S.-I. (2017). A unified approach to interpreting model predictions. In *Advances in neural information processing systems* (pp. 4765–4774).
46. Makhzani, A., Shlens, J., Jaitly, N., Goodfellow, I., & Frey, B. (2015). Adversarial autoencoders. arXiv preprint arXiv:1511.05644.
47. Malgieri, G., & Comandé, G. (2017). Why a right to legibility of automated decision-making exists in the General Data Protection Regulation. *International Data Privacy Law, 7*(4), 243–265.
48. Martens, D., Baesens, B., Van Gestel, T., & Vanthienen, J. (2007). Comprehensible credit scoring models using rule extraction from support vector machines. *European Journal of Operational Research, 183*(3), 1466–1476.
49. Miller, T. (2019). Explanation in artificial intelligence: Insights from the social sciences. *Artificial Intelligence, 267*, 1–38.
50. Molnar, C. (2020). *Interpretable machine learning*. Lulu.com
51. Murphy, P. M., & Pazzani, M. J. (1991). Id2-of-3: Constructive induction of m-of-n concepts for discriminators in decision trees. In *Machine learning proceedings 1991* (pp. 183–187). Elsevier.
52. Naretto, F., Pellungrini, R., Monreale, A., Nardini, F. M., & Musolesi, M. (2020). Predicting and explaining privacy risk exposure in mobility data. In *International Conference on Discovery Science* (pp. 403–418). Springer.
53. Oriol, J. D. V., Vallejo, E. E., Estrada, K., Peña, J. G. T., Initiative, A. D. N., et al. (2019). Benchmarking machine learning models for late-onset Alzheimer's disease prediction from genomic data. *BMC Bioinformatics, 20*(1), 1–17.
54. Pasquale, F. (2015). *The black box society*. Harvard University Press.
55. Pedreschi, D., Giannotti, F., Guidotti, R., Monreale, A., Ruggieri, S., & Turini, F. (2019). Meaningful explanations of black box AI decision systems. In *Proceedings of the AAAI Conference on Artificial Intelligence* (Vol. 33, pp. 9780–9784).
56. Pedreshi, D., Ruggieri, S., & Turini, F. (2008). Discrimination-aware data mining. In *Proceedings of the 14th ACM SIGKDD International Conference on Knowledge Discovery and Data Mining* (pp. 560–568).
57. Quinlan, J. R. (1993). *C4.5: Programs for machine learning*. Elsevier.
58. Ribeiro, M. T., Singh, S., & Guestrin, C. (2016). Why should I trust you? Explaining the predictions of any classifier. In *Proceedings of the 22nd ACM SIGKDD International Conference on Knowledge Discovery and Data Mining* (pp. 1135–1144). ACM.
59. Ribeiro, M. T., Singh, S., & Guestrin, C. (2018). Anchors: High-precision model-agnostic explanations. In *Thirty-Second AAAI Conference on Artificial Intelligence*.
60. Romei, A., & Ruggieri, S. (2014). A multidisciplinary survey on discrimination analysis. *Knowledge Engineering Review, 29*(5), 582–638.
61. Rudin, C. (2019). Stop explaining black box machine learning models for high stakes decisions and use interpretable models instead. *Nature Machine Intelligence, 1*(5), 206–215.

62. Rudin, C., & Radin, J. (2019). Why are we using black box models in AI when we don't need to? A lesson from an explainable AI competition. *Harvard Data Science Review, 1*(2). https://hdsr.mitpress.mit.edu/pub/f9kuryi8

63. Setzu, M., Guidotti, R., Monreale, A., & Turini, F. (2019). Global explanations with local scoring. In *Joint European Conference on Machine Learning and Knowledge Discovery in Databases* (pp. 159–171). Springer.

64. Shapley, L. S. (1953). A value for n-person games. *Contributions to the Theory of Games, 2*(28), 307–317.

65. Shokri, R., Strobel, M., & Zick, Y. (2019). Privacy risks of explaining machine learning models. CoRR, abs/1907.00164.

66. Shrikumar, A., Greenside, P., Shcherbina, A., & Kundaje, A. (2016). Not just a black box: Learning important features through propagating activation differences. arXiv preprint arXiv:1605.01713.

67. Simonyan, K., Vedaldi, A., & Zisserman, A. (2013). Deep inside convolutional networks: Visualising image classification models and saliency maps. arXiv preprint arXiv:1312.6034.

68. Sokol, K., & Flach, P. A. (2020). Explainability fact sheets: A framework for systematic assessment of explainable approaches. In *FAT* '20: Conference on Fairness, Accountability, and Transparency, Barcelona, Spain, January 27–30, 2020* (pp. 56–67). ACM.

69. Sundararajan, M., Taly, A., & Yan, Q. (2017). Axiomatic attribution for deep networks. arXiv preprint arXiv:1703.01365.

70. Swapna, G., Vinayakumar, R., & Soman, K. (2018). Diabetes detection using deep learning algorithms. *ICT Express, 4*(4), 243–246.

71. Tan, P.-N. et al. (2006). *Introduction to data mining*. Pearson Education India.

72. Ustun, B., Spangher, A., & Liu, Y. (2019). Actionable recourse in linear classification. In *Proceedings of the Conference on Fairness, Accountability, and Transparency* (pp. 10–19).

73. Wachter, S., Mittelstadt, B., & Floridi, L. (2017). Why a right to explanation of automated decision-making does not exist in the General Data Protection Regulation. *International Data Privacy Law, 7*(2), 76–99.

74. Wachter, S., Mittelstadt, B., & Russell, C. (2017). Counterfactual explanations without opening the black box: Automated decisions and the GDPR. *HJLT, 31*, 841.

75. Wang, F., & Rudin, C. (2015). Falling rule lists. In *Artificial intelligence and statistics* (pp. 1013–1022).

76. Yin, X., & Han, J. (2003). CPAR: Classification based on predictive association rules. In *Proceedings of the 2003 SIAM International Conference on Data Mining* (pp. 331–335). SIAM.

77. Zeiler, M. D., & Fergus, R. (2014). Visualizing and understanding convolutional networks. In *European Conference on Computer Vision* (pp. 818–833). Springer.

Chapter 3
Science of Data: A New Ladder for Causation

Usef Faghihi, Sioui Maldonado Bouchard, and Ismail Biskri

3.1 Introduction

Deep learning (DL) algorithms are at the center of weak **artificial** intelligence (AI) technology. Among others, they are used for traffic forecasting [1, 2], sentiment analysis [3], and face detections [4].

Although much effort has been invested in integrating reasoning and explainability in DL algorithms recently [5], this AI continues to lack in both explainability (we do not understand how any particular DL algorithm achieves its tasks) and reasoning (it cannot reason by finding causes of events or situations), as Faghihi [6, 7] and Pearl [8] have pointed out. For the purpose of this chapter, our working definition of causality is the connections (elements) that explain result.

DL algorithms need large amounts of data to learn [8], and without explainability and reasoning, one cannot decide why and how DL algorithms make decisions. This allows no room for improvement in AI [5]. On the other hand, inferential and fuzzy logic are very effective for reasoning [6, 8]. However, they cannot learn and most of the examples given by logicians work perfectly only in a laboratory setting [7].

Spontaneous human reasoning is autonomous yet in constant interactions with its environment. Humans use different types of memory, learning, and logic in their interactions with their environment and among themselves for reasoning. Cognitive architectures simulate the functionality of the human mind [9–11]. They use different types of memories for learning and reasoning. They are equipped with the detailed implementation of the theory of mind such as attention [12].

U. Faghihi · I. Biskri (✉)
University of Québec at Trois-Rivières, Trois-Rivières, QC, Canada
e-mail: usef.faghihi@uqtr.ca; ismail.biskri@uqtr.ca

S. M. Bouchard
University of Montreal, Montreal, QC, Canada
e-mail: sioui.maldonado@umontreal.ca

© The Author(s), under exclusive license to Springer Nature Switzerland AG 2021
M. Sayed-Mouchaweh (ed.), *Explainable AI Within the Digital Transformation and Cyber Physical Systems*, https://doi.org/10.1007/978-3-030-76409-8_3

However, current cognitive architectures are not equipped with neural networks, so they cannot perform detailed and precise image detection. We think that this is time to integrate neural networks into cognitive architectures to create a more transparent (read *explainable*) AI that can reason.

In the following sections, we briefly review cognitive architectures, inferential logic, probabilistic fuzzy logic (PFL), and neural networks. We then give a brief overview of graph neural networks. We finally suggest a cognitive architecture that, in addition to its current machine learning algorithms, will be equipped with Probabilistic Fuzzy logic and Graph Neural Networks for reasoning.

3.2 Related Works

3.2.1 Cognitive Architectures

Humans use different types of memory, learning, and logic in their interactions with their environment and among themselves for reasoning. Cognitive architectures [10, 13–16] implement a theory of mind according to the neuroscience theories of cognition [17]. The most famous cognitive architectures are Newell's Soar architecture [16, 18], Anderson's ACT-R architecture [11], Sun's CLARION architecture [10], and Franklin's LIDA architecture [13]. These cognitive architectures each have their strengths and weaknesses when implementing and explaining a theory of mind [14]. Current cognitive architectures are limited to meta-reasoning. Theoretically, cognitive architectures are capable of reasoning about the relationship between two persons in a photo. However, they cannot compete with neural networks such as Attentive Graph Convolutional Neural Networks for detecting objects in a photo. Conversely, Attentive Graph Convolutional Neural Networks detect objects in a photo but have a limited ability to reason about their relationship [19].

In 2018, Canadian governmental research agencies invested $125 million in AI [20]. Part of these funds has gone to the creation of attentional neural networks based on cognitive neuroscience theory [21–23]; yet all the while, and for more than a decade, cognitive architectures have already had working attentional mechanisms [12].

One could argue that it would be great to integrate DL algorithms *with* cognitive architectures [24]. In addition to DL algorithms, we suggest that cognitive architectures need to be equipped with nonclassical logic such as inferential logic and/or probabilistic fuzzy logic.

3.2.2 Inferential Logic

Pearl proposes inferential logic as the ideal framework for reasoning [8]. He suggests three levels for causality: (1) **association**—to identify interrelated phenomena;

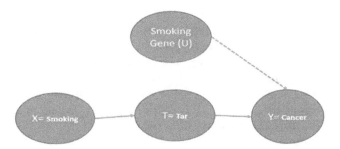

Fig. 3.1 $P(y|x) \neq P(y|do(x))$. The probability of a patient having lung cancer $Y = y$, given that we intervene and ask the person to smoke a pack of cigarette per day (set the value of X to x) and subsequently observe what happens [8]

(2) **intervention**—to predict the consequences of an action (i.e., how does my life expectancy change if I became a vegetarian?); (3) **counterfactual**—to reason about hypothetical situations and possible outcomes (i.e., would my grandfather still be alive if he had not smoked?). However, Pearl's approach to causation does not allow to simulate real-life problems nor does it allow for reasoning in cases of degrees of certainty. With Pearl's approach, to simulate real-life problems, we need to use Directed Acyclic Graphs (DAGs). In DAGs, nodes cannot perform two-way communications. However, real-life problems almost always require two-way communication. For instance, finding a faulty node in a network where every node sends and receives data from/to other nodes. Another problem with Pearl's approach is the *do* operators which cut the relation between two nodes in DAGs. For instance, suppose we have four nodes in a DAG: Cancer Gene, Smoke, Tar, Cancer (Fig. 3.1). In this example, suppose the Cancer Gene node has an arrow to the Smoke node and the Cancer node but no arrow to the Tar node. Further, the Smoke node has a direct arrow to the Tar node, and the Tar node has a direct arrow to the Cancer node. The *do* operator cuts the arrow between the Cancer Gene node and Smoke node and fixes the Smoke node's value (Fig. 3.1).

This is akin to forcing a smoker to smoke a packet of cigarettes a day and then observing whether the Cancer node's value changes [8]. However, Pearl's approach to causality cannot answer gradient questions such as: given that you smoke a little, what is the probability that you have cancer of a certain severity? [6, 7]. An alternative to inferential logic is probabilistic fuzzy logic.

3.2.3   Probabilistic Fuzzy Logic (PFL)

PFL, on the other hand, excels at reasoning with degrees of certainty and in real-life problems [25]. Importantly, this allows for degrees of dependency and membership. In PFL, Zadeh [26] proposes that a set of elements always has a degree of membership between [0,1]. PFL processes three types of uncertainty:

randomness, probabilistic uncertainty, and fuzziness [7]. PFL can both manage the uncertainty of our knowledge (by the use of probabilities) and the vagueness inherent to the world's complexity (by data fuzzification) [25]. PFL has been used to solve many engineering problems, such as security intrusion detection [27, 28] and finding the causes of events [6, 7]. Faghihi et al. in [6, 7] showed that PFL outperforms inferential logic. However, PFL cannot learn by itself and needs experts to define intervals before applying fuzzification [6, 7]. Among others, one can integrate PFL with Neural Networks for learning.

3.2.4 Neural Networks (NN)

Neural Networks (NN) outperform humans in specific tasks such as Google AlphaGo [29]. However, they are mostly single-task tools and behave differently when the nature of data changes [30]. We can consider NNs as predominantly **Microscopic** or **Macroscopic**; they each have their advantages, but also their limitations. In the next two subsections, we will very briefly explain them.

3.2.4.1 Microscopic Neural Network (NN)

Microscopic NNs are considered Structural Causal Models and seek to be interpretable. The problem with most of the Microscopic NN is that since most of the nodes in NNs are interrelated, a small change in any node's weight results in changes in the weights of almost all the nodes. Implementing causality with such NNs is a challenge [31]. Given an input, we need to understand why NNs make particular errors and identify the nodes responsible for the output errors. To do so, some researchers use game theories such as Shapley [32–34]. However, given a specific data type, these techniques assign importance to certain nodes or features that may never be used by the technique [30].

Furthermore, given a NN, some researchers use the Dropout [35, 36] technique to find the most influential neurons that cause NN's output. However, using the Dropout method to find the cause(s) of an event in a NN will result in a combinatorial explosion [31]. In [31], the authors tried to implement a very simple causality model using an inner and outer loop. The inner loop captures the local distribution changes for a specific node given its neighbors. The outer loop learns the model's meta-parameters [31]. However, although in [31] the authors claimed that their model captures intervention *a la Pearl* [8], the authors failed to implement a real-life problem.

3.2.4.2 Macroscopic Neural Network (NN)

The idea for **Macroscopic NN** comes from neuroscience evidence of brain activity when faced with tasks or images [37]. That is, for a given task, one or more regions of the brain are activated [38]. **Macroscopic NN** AI efforts are undertaken to imitate cognitive architectures, such as adding attention to neural networks. However, efforts may be better served by restructuring neural nets to use them *within* cognitive architecture [24]. Recently, Graph Neural Networks[1] (GNNs) got much attention in the AI community [39–44]. In the next section, we will briefly introduce GNNs. We then explain how one can integrate inferential or fuzzy logics and deep learning algorithms with cognitive architectures.

3.2.4.3 Graph Neural Networks

Many real-world problems can be simulated with graphs [45–47]. For instance, predicting node relationships and their behaviors on Facebook or Pinterest. A graph represents a data structure consisting of the nodes and edges (Fig. 3.2). In a Directed Graph, there is an order between the pair of the nodes and the edges, whereas in an Undirected Graph, there is no specific direction between nodes. One can represent a graph using an adjacency matrix. Graphs are considered as non-Euclidian geometry, which means the shortest path between two nodes may not be represented with a straight line [45]. This makes graph interpretation rather challenging.

Another challenge with the graphs is their permutation invariance/equivariance feature. Two graphs may seem visually different but have the same adjacency matrix representation [44]. This also makes node classifications and nodes' behavior predictions difficult.

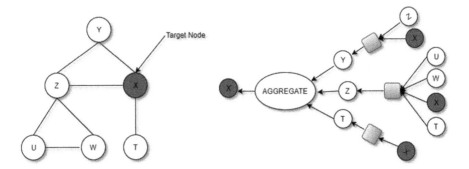

Fig. 3.2 How GNNs process graph nodes [45]. The graph structure is on the left. The aggregation between the graph's nodes is shown on the right

---

[1]A graph represents a data structure consisting of the nodes and edges.

To overcome these problems when analyzing graphs, researchers use Graph Neural Networks (GNNs) [39]. GNNs' nodes are connectionist models. They are defined according to their relationships with their neighbors and connections between them [39, 40]. As opposed to neural networks, GNNs' architecture represents the state of every node according to the messages they receive from their neighborhood nodes. In GNNs, each node contains its unique features, the edges connecting it to other nodes in the graph, information regarding the state, and the features of the neighboring nodes. Being highly connected, some nodes have more connections to other nodes in the graph. This makes them more important than other nodes in the graph in the learning and prediction steps [41].

To overcome this problem, among others, researchers use the following architectures [41, 42]: (1) **Recurrent graph neural networks**, where every node in the graphs has a connection to itself which makes them recurrent. Nodes simultaneously receive messages from their neighbors until the graph converges to a stable state. The recurrent nature of this graph prevents any data loss in the nodes; (2) **Convolutional graph neural networks (ConvGNNs),** where every node aggregates its information with the incoming information from its neighbors. ConvGNNs usually use many hidden layers to classify nodes or graphs; (3) **Graph autoencoders (GAEs),** where the model encodes/graphs into a latent vector space. The model then decodes or reconstructs the graph using encoded hidden space. GAE is an unsupervised model; (4) **Spatial-temporal graph neural networks (STGNNs)**, where the model encodes spatial and temporal features of the real-life objects into graph's nodes; for instance, human action recognition [43].

To do reasoning with graph neural networks (GNNs), among others, one needs to use asynchronous message passing between nodes. This is important because for instance in causality, the ordering of the nodes in GNNs matters [48]. Furthermore, instead of average or sum message passing methods for each node, one needs to add logical rules similar to [6, 7] to find the causes in GNNs. Another alternative for reasoning with neural networks is hybrid Neural Networks. Unlike GNNs, Hybrid Neural Networks integrate different tools for reasoning. As such, they are similar to what we suggest for reasoning and causality in this study.

3.2.4.4 Hybrid Neural Networks for Reasoning

Among others, the authors in [5, 49] suggest a hybrid architecture consisting of a Graph Neural Network (GNN) and two Multi-Layer Perceptron (MLP) acting as its modules, applied to pairs of objects, capable of solving some specific type of relational reasoning such as dynamic programming. MLPs are great in learning and predictions of a single object but cannot generalize facing aggregated nodes in GNNs. However, the authors did not discuss how to apply counterfactual reasoning as explained by Faghihi et al. [6, 7] or how to perform reasoning when we are facing unknown structures.

As we can see here, GNNs alone are not sufficient for reasoning [5, 14, 24, 44, 46]. We need to use different techniques and algorithms to create a successful AI tool for reasoning.

We suggest our combining cognitive architectures, PFL, and GNNs as a theoretical model for reasoning.

3.3   Cognitive Architecture Equipped with PFL and GNNs

The reality is that merely integrating PFLs with NNs or GNNs will not be a successful combination for reasoning [24, 50, 51]. Similar to humans, to reason, cognitive architectures need to use different types of memory, neural networks, and/or GNNs, inferential/fuzzy logic, and machine learning algorithms such as breadth-first search (BFS) in order for their reasoning processes to be explainable [24, 52].

Among different cognitive architectures, we select here **Learning Intelligent Distributed Agents (LIDA)** [13, 53]. LIDA is based on the Global Workspace Theory (GWT). GWT is one of the most widely accepted theories on how the mind works in the field of neuroscience [13, 54]. LIDA's architecture, equipped with different modules such as artificial motivation [13], makes decisions and learns using its cognitive cycles (Fig. 3.3). A cognitive cycle starts by perception/understanding [55] of its environment, an attending phase [12], and an action selection/learning phase [12, 56, 57]. The full explanation of LIDA's architecture is out of the scope of this chapter.

Roughly speaking, each module may have one or more codelets. Codelets are small pieces of code each conceived to do a specific task such as detecting a car. In every cognitive cycle, after sensing the environment, once LIDA's perceptual nodes are activated, they go to a temporary place called Current Situational Model (CSM) (Fig. 3.3). LIDA's Attention module decides what portion of the represented nodes in CSM is most in need of LIDA's Attention; it determines what nodes are most salient [6, 13]. It then creates a coalition(s) of codelets from the most activated codelets. This portion of codelets (the graphs in NN language), considered the most important/appropriate to the current situation, is broadcast to the rest of the system, making it the current contents of consciousness to reasoning and action selection [9].

Although they are not the same, codelets can be seen as nodes or combinations of the nodes in GNNs—*one can replace codelets with neurons*. In the following, we will briefly explain how to integrate association/causal PFL rules with GNNs into LIDA's **Attention** and **Reasoning** modules. With this integration, explainable and detailed AI becomes possible.

Attention Module: Similarly to [5], to create coalitions of neurons in CSM, LIDA's classic Attention module performs as follows: for every neuron activated in perception and having arrived in the Current Situational Model (CSM), the Attention module updates the states of the node with the activation of its neighbors (similar to

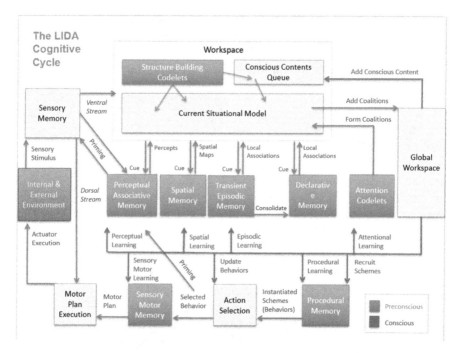

Fig. 3.3 LIDA's cognitive cycle

graphs [5] receiving messages from other nodes). In our model, we also use LIDA, but add a fuzzification step as explained in [6, 7].

Similarly to [31], but using probabilistic Fuzzy logic rules [6, 7], for all the nodes pointing to node *I* in CSM, our model uses fuzzification of association rules as follow:

$$\text{fuzz } x_i^l = \text{Fuzzification}\left(x_{i \leftarrow j}^{(l-1)}, a_{i \leftarrow j}, \quad \text{where } j \neq i \right) : a_i, \ a_j \in N,$$

l correspond to graph's layer, x_i is a vector for each node v_i, $a_{i \leftarrow j}$ is the activation of node *i* in CSM.

Using the above strategy, our modified Attention module (**AM**) creates a graph for every node (i.e., node *x* in Fig. 3.2) in the CSM. To compute the most activated graph in the CSM, our modified AM then compares outputs for the fuzzification of the nodes. Our modified AM then finds the graph with the most activated nodes in CSM and sends it for conscious competition.

Upon receiving the graph, in our model, LIDA's conscious module broadcasts it to all LIDA's modules. This may result in LIDA's modules adding extra nodes to the broadcast graph (for instance LIDA's Episodic memory can add nodes to the graph). This alters the graph previously created by the AM. The modified graph is broadcast again by the conscious module to all other LIDA's modules until a decision is made by LIDA.

Reasoning module (RM): One of LIDA's modules that receives the graph broadcast by the consciousness module is the Reasoning module (**RM**). To find the causal nodes in the graph, in our model, LIDA's RM uses PFL's defuzzification techniques, such as applying min max rules from [6, 7] to the graph's nodes.

To do so, LIDA's module calculates every node's influence on its neighbors as follows:

1. To calculate how a specific node s influences node i, the RM applies fuzzy dual **max** rule *a la* [6, 7]: Suppose node $s \in (1, \ldots, s, \ldots j)$ is one of the incoming nodes to a node i. Similarly to [31] but using probabilistic Fuzzy logic rules [6, 7],

$$\text{defuzz max } x_s^l = \text{Dual}\left(\left(\max\left(1 - s_{i \leftarrow s}^{(l-1)}, a_i\right) : \quad \text{where} \quad s \neq i, \ v_i \in N_i^*\right)\right),$$

   a_i is the activation of node i, l corresponds to graph's layer.
2. Step 1 is repeated for the incoming node s to the node i; except this time, our model uses the Dual **minimum** rule to minimize the influence of node s:

$$\text{defuzz min } x_s^l \text{Dual}\left(\left(\min\left(1 - s_{i \leftarrow s}^{(l-1)}, a_i\right) : \quad \text{where} \quad s \neq j, \ v_j \in N_i^*\right), v_i\right),$$

   a_i is the activation of node i.
3. Similarly to [5], in our model, LIDA's RM uses MLPs, to compute and learn the discrepancies between applying Dual min and max for the incoming node s to node i. If there is a significant difference between the graph's output for Dual min and max, RM then tags node s as a possible cause. Learning occurs by using MLP as follow: outputs:

$$h_i^{(l)} = \sum_{t \in J} \text{MLP}_1^{(l)}\left(\text{fuzz } x_i^{l-1}, \text{fuzz } x_j^l\right), \quad h_i = \text{MLP}_2\left(\sum_{i \in J} h_i^{(L)}\right),$$

   where h_i is the graphs' output.
4. RM applies steps 1–3 to all nodes in the graph received by RM and tags possible causal nodes. RM then sends the graph again to compete for consciousness and action selection. We are aware, as pointed out by [5], that using MPLs as modules for GNNs may lead to the overparameterized[2] problem. This can be solved using simple polynomial functions such as gradient descent [5].

In the above Attention and Reasoning mechanisms, we briefly explained a theoretical framework in which by combining different types of logic, graph neural networks can be integrated into a cognitive architecture and bring more clarity on how a system makes decisions. This is in part possible because LIDA architecture allows a detailed implementation of all its modules [12].

---

[2]That is, when the number of parameters in a learned network has more parameters than the target.

That is, for instance, if a node is tagged as a possible cause by RM, we can know why RM tagged that node as a cause. This, if needed, gives us the possibility of changing or modifying the graph selection mechanism in LIDA's AM. Furthermore, thanks to logic, one can interpret why a specific node in a graph is tagged as a possible cause and why a specific graph is selected by AM to broadcast for conscious competition. Using, for instance, reinforcement learning, one can correct LIDA's decision-making module's mistakes.

The above framework is a clear step toward creating a reasoning—and consequently explainable—AI tool.

3.4 Conclusion

Recently, researchers in the domain of artificial intelligence have sought to use neuroscience evidence of how the brain works to create more sophisticated neural networks, such as attentional neural networks [21, 58]. Yet, all the while, cognitive architectures are already equipped with, for instance, attention and attentional learning modules [12, 15].

In this chapter, we suggested a theoretical framework to create an explainable AI tool that is capable of reasoning. To do so, among others, the AI tool needs to be equipped with different types of logic, Neural Networks, and Graph Neural Networks integrated with cognitive architectures.

References

1. Li, Y., Yu, R., Shahabi, C., & Liu, Y. (2017). Diffusion convolutional recurrent neural network: Data-driven traffic forecasting. *arXiv preprint arXiv:1707.01926.*
2. Yu, B., Yin, H., & Zhu, Z. (2017). Spatio-temporal graph convolutional networks: A deep learning framework for traffic forecasting. *arXiv preprint arXiv:1709.04875.*
3. Baziotis, C., Pelekis, N., & Doulkeridis, C. (2017). Data stories at SemEval-2017 task 4: Deep LSTM with attention for message-level and topic-based sentiment analysis. In *Proceedings of the 11th international workshop on semantic evaluation (SemEval-2017)* (pp. 747–754).
4. Ding, X., Raziei, Z., Larson, E. C., Olinick, E. V., Krueger, P., & Hahsler, M. (2020). Swapped face detection using deep learning and subjective assessment. *EURASIP Journal on Information Security, 2020,* 1–12.
5. Xu, K., Li, J., Zhang, M., Du, S. S., Kawarabayashi, K. I., & Jegelka, S. (2019). What can neural networks reason about? *arXiv preprint arXiv:1905.13211.*
6. Faghihi, U., Robert, S., Poirier, P., & Barkaoui, Y. (2020). From Association to Reasoning, an Alternative to Pearl's Causal Reasoning. In *Proceedings of AAAI-FLAIRS 2020.*
7. Robert, S., Faghihi, U., Barkaoui, Y., & Ghazzali, N. (2021). Causality in probabilistic fuzzy logic and alternative causes as fuzzy duals. In *ICCCI 2020: Advances in computational collective intelligence.*
8. Pearl, J., & Mackenzie, D. (2018). *The book of why: The new science of cause and effect.* Basic Books.

9. Faghihi, U., Estey, C., McCall, R., & Franklin, S. (2015). A cognitive model fleshes out Kahneman's fast and slow systems. *Biologically Inspired Cognitive Architectures, 11*, 38–52.
10. Sun, R. (2020). Potential of full human–machine symbiosis through truly intelligent cognitive systems. *AI & Society, 35*, 17–28.
11. Bono, A., Augello, A., Pilato, G., Vella, F., & Gaglio, S. (2020). An ACT-R based humanoid social robot to manage storytelling activities. *Robotics, 9*, 25.
12. Faghihi, U., McCall, R., & Franklin, S. (2012). A computational model of attentional learning in a cognitive agent. *Biologically Inspired Cognitive Architectures, 2*, 25–36.
13. McCall, R. J., Franklin, S., Faghihi, U., Snaider, J., & Kugele, S. (2020). Artificial motivation for cognitive software agents. *Journal of Artificial General Intelligence, 11*, 38–69.
14. Faghihi, U., & Franklin, S. (2012). The LIDA model as a foundational architecture for AGI. In *Theoretical foundations of artificial general intelligence* (pp. 103–121). Springer.
15. Anderson, J. R., Matessa, M., & Lebiere, C. (1997). ACT-R: A theory of higher level cognition and its relation to visual attention. *Human Computer Interaction, 12*, 439–462.
16. Laird, J. E. (2012). *The Soar cognitive architecture*. MIT Press.
17. Lieto, A., Bhatt, M., Oltramari, A., & Vernon, D. (2018). *The role of cognitive architectures in general artificial intelligence*. Elsevier.
18. Laird, J. E., Newell, A., & Rosenbloom, P. S. (1986). *SOAR: An architecture for general intelligence*. Stanford University, Department of Computer Science.
19. Li, S., Tang, M., Zhang, J., & Jiang, L. (2020). Attentive gated graph neural network for image scene graph generation. *Symmetry, 12*, 511.
20. Williams, M. A. (2019). The Artificial Intelligence race: Will Australia lead or lose? In *Journal and Proceedings of the Royal Society of New South Wales* (p. 105). Royal Society of New South Wales.
21. Veličković, P., Cucurull, G., Casanova, A., Romero, A., Lio, P., & Bengio, Y. (2017). Graph attention networks. *arXiv preprint arXiv:1710.10903*.
22. Choi, H., Cho, K., & Bengio, Y. (2018). Fine-grained attention mechanism for neural machine translation. *Neurocomputing, 284*, 171–176.
23. Qu, M., Bengio, Y., & Tang, J. (2019). GMNN: Graph Markov neural networks. *arXiv preprint arXiv:1905.06214*.
24. Perconti, P., & Plebe, A. (2020). Deep learning and cognitive science. *Cognition, 203*, 104365.
25. Yager, R. R., & Zadeh, L. A. (2012). *An introduction to fuzzy logic applications in intelligent systems*. Springer Science & Business Media.
26. Zadeh, L. A., Klir, G. J., & Yuan, B. (1996). *Fuzzy sets, fuzzy logic, and fuzzy systems: Selected papers*. World Scientific.
27. Zhao, D.-M., Wang, J.-H., Wu, J., & Ma, J.-F. (2005). Using fuzzy logic and entropy theory to risk assessment of the information security. In *2005 International Conference on Machine Learning and Cybernetics* (pp. 2448–2453). IEEE.
28. Cheng, P.-C., Rohatgi, P., Keser, C., Karger, P. A., Wagner, G. M., & Reninger, A. S. (2007). Fuzzy multi-level security: An experiment on quantified risk-adaptive access control. In *2007 IEEE Symposium on Security and Privacy (SP'07)* (pp. 222–230). IEEE.
29. Granter, S. R., Beck, A. H., & Papke, D. J., Jr. (2017). AlphaGo, deep learning, and the future of the human microscopist. *Archives of Pathology & Laboratory Medicine, 141*, 619–621.
30. Chen, H., Janizek, J. D., Lundberg, S., & Lee, S. -I. (2020). True to the model or true to the data? *arXiv preprint arXiv:2006.16234*.
31. Ke, N. R., Bilaniuk, O., Goyal, A., Bauer, S., Larochelle, H., Pal, C., & Bengio, Y. (2019). Learning neural causal models from unknown interventions. *arXiv preprint arXiv:1910.01075*.
32. Shapley, L. S. (1953). A value for n-person games. *Contributions to the Theory of Games, 2*, 307–317.
33. Chattopadhyay, A., Manupriya, P., Sarkar, A., & Balasubramanian, V. N. (2019). Neural network attributions: A causal perspective. *arXiv preprint arXiv:1902.02302*.
34. Janzing, D., Minorics, L., & Blöbaum, P. (2020). Feature relevance quantification in explainable AI: A causal problem. In *International Conference on artificial intelligence and statistics* (pp. 2907–2916).

35. Malach, E., Yehudai, G., Shalev-Shwartz, S., & Shamir, O. (2020). Proving the Lottery Ticket hypothesis: Pruning is all you need. *arXiv preprint arXiv:2002.00585.*
36. Gal, Y., & Ghahramani, Z. (2016). Dropout as a Bayesian approximation: Representing model uncertainty in deep learning. In *International conference on machine learning* (pp. 1050–1059).
37. Miyoshi, T., Tanioka, K., Yamamoto, S., Yadohisa, H., Hiroyasu, T., & Hiwa, S. (2020). Revealing changes in brain functional networks caused by focused-attention meditation using Tucker3 clustering. *Frontiers in Human Neuroscience, 13,* 473.
38. Christoff, K., Prabhakaran, V., Dorfman, J., Zhao, Z., Kroger, J. K., Holyoak, K. J., & Gabrieli, J. D. (2001). Rostrolateral prefrontal cortex involvement in relational integration during reasoning. *NeuroImage, 14,* 1136–1149.
39. Scarselli, F., Gori, M., Tsoi, A. C., Hagenbuchner, M., & Monfardini, G. (2008). The graph neural network model. *IEEE Transactions on Neural Networks, 20,* 61–80.
40. Cao, S., Lu, W., & Xu, Q. (2016). Deep neural networks for learning graph representations. In *Thirtieth AAAI conference on artificial intelligence.*
41. Ruiz, L., Gama, F., & Ribeiro, A. (2020). Gated graph recurrent neural networks. *arXiv preprint arXiv:2002.01038.*
42. Wu, Z., Pan, S., Chen, F., Long, G., Zhang, C., & Philip, S. Y. (2020). A comprehensive survey on graph neural networks. In *IEEE transactions on neural networks and learning systems.*
43. Yan, S., Xiong, Y., & Lin, D. (2018). Spatial temporal graph convolutional networks for skeleton-based action recognition. In *Thirty-second AAAI conference on artificial intelligence.*
44. Loukas, A. (2019). What graph neural networks cannot learn: depth vs width. *arXiv preprint arXiv:1907.03199.*
45. Hamilton, W. L., Ying, R., & Leskovec, J. (2017). Representation learning on graphs: Methods and applications. *arXiv preprint arXiv:1709.05584.*
46. Xu, K., Hu, W., Leskovec, J., & Jegelka, S. (2018) How powerful are graph neural networks? *arXiv preprint arXiv:1810.00826.*
47. You, J., Ying, R., Ren, X., Hamilton, W. L., & Leskovec, J. (2018). GraphRNN: Generating realistic graphs with deep auto-regressive models. *arXiv preprint arXiv:1802.08773.*
48. Zhang, M., Jiang, S., Cui, Z., Garnett, R., & Chen, Y. (2019). D-VAE: A variational autoencoder for directed acyclic graphs. In *Advances in Neural Information Processing Systems* (pp. 1588–1600).
49. Battaglia, P. W., Hamrick, J. B., Bapst, V., Sanchez-Gonzalez, A., Zambaldi, V., Malinowski, M., Tacchetti, A., Raposo, D., Santoro, A., & Faulkner, R. (2018). Relational inductive biases, deep learning, and graph networks. *arXiv preprint arXiv:1806.01261.*
50. Subagdja, B., & Tan, A.-H. (2015). Neural modeling of sequential inferences and learning over episodic memory. *Neurocomputing, 161,* 229–242.
51. CHIA, H. W.-K., & TAN, C.-L. (2001). Neural logic network learning using genetic programming. *International Journal of Computational Intelligence and Applications, 1,* 357–368.
52. Yoo, A., Chow, E., Henderson, K., McLendon, W., Hendrickson, B., & Catalyurek, U. (2005). A scalable distributed parallel breadth-first search algorithm on BlueGene/L. In *SC'05: Proceedings of the 2005 ACM/IEEE Conference on Supercomputing* (pp. 25–25). IEEE.
53. Faghihi, U., Maldonado-Bouchard, S., & Incayawar, M. (2020). In M. Incayawar & S. Maldonado-Bouchard (Eds.), *Taming artificial intelligence in psychiatry and pain medicine— Promises and challenges.* Oxford University Press.
54. Grinde, B., & Stewart, L. (2020). A global workspace, evolution-based model of the effect of psychedelics on consciousness. In *Psychology of consciousness: theory, research, and practice.*
55. Ryan, K., Agrawal, P., & Franklin, S. (2020). The pattern theory of self in artificial general intelligence: A theoretical framework for modeling self in biologically inspired cognitive architectures. *Cognitive Systems Research, 62,* 44–56.
56. D'Mello, S. K., Ramamurthy, U., Negatu, A., & Franklin, S. (2006). A procedural learning mechanism for novel skill acquisition. In *Workshop on motor development: Proceeding of adaptation in Artificial And Biological Systems, AISB'06.* Citeseer.

57. Dong, D., & Franklin, S. (2015). Modeling sensorimotor learning in LIDA using a dynamic learning rate. *Biologically Inspired Cognitive Architectures, 14*, 1–9.
58. Chi, L., Yuan, Z., Mu, Y., & Wang, C. (2020). Non-local neural networks with grouped bilinear attentional transforms. In *Proceedings of the IEEE/CVF Conference on computer vision and pattern recognition* (pp. 11804–11813).

Chapter 4
Explainable Artificial Intelligence for Predictive Analytics on Customer Turnover: A User-Friendly Interface for Non-expert Users

Joglas Souza and Carson K. Leung ⓘ

4.1 Introduction

Nowadays, data—which can be considered as valuable as the "new oil"—are everywhere. Artificial intelligence [3] (especially, deep learning [13, 33] within the area of machine learning) techniques, together with data mining [9, 17, 31] and big data science [27, 29, 30] solutions, have become critical for decision-making mechanisms in numerous real-life applications and service domains. Examples include critical systems, cyber-physical systems (CPS) [7], digital transformation [8], e-government [4], finance [35], healthcare [42], industry 4.0 [43], justice [5], predictive maintenance [38], smart energy management [6], smart factory [51], and smart grids [19]. Analyses of big data from these domains—from census analysis [11] to social network analysis [18, 24]; from music analytics [16, 53] to movie analytics [20, 34]; from disease analytics [10, 45, 48] to transportation/urban analytics [32, 36]—have led to valuable knowledge and information. However, recommendations made by these techniques and analyses, as well as their logical reasoning behind these recommendation decisions, are often not easy to be comprehended by humans. Good-quality explanations are needed for humans to trust and collaborate with such intelligent systems [1]. The more impact a machine learning decision has on customer's or people's lives, the higher is the necessity for its explanation [40]. The need for explanation is not only a demand from end users but also a regulatory requirement in some countries. For instance, the *General Data Protection Regulation* (*GDPR*) that has been implemented in European countries recently states that customers have the right of explanation for decisions made through automated systems [21]. Explanations are also a means of detecting model bias as it can reveal when the model is making decisions

J. Souza · C. K. Leung (✉)
University of Manitoba, Winnipeg, MB, Canada
e-mail: kleung@cs.umanitoba.ca

© The Author(s), under exclusive license to Springer Nature Switzerland AG 2021   47
M. Sayed-Mouchaweh (ed.), *Explainable AI Within the Digital Transformation and Cyber Physical Systems*, https://doi.org/10.1007/978-3-030-76409-8_4

based on incorrect assumptions. The enhancement of models also relies on how comprehensive the model outcome is to propose the appropriate improvements [1]. *EXplainable Artificial Intelligence (XAI)* [12, 15], the research area that studies how to make models transparent and explainable, is now in the spotlight for keeping the adoption of machine learning growing.

In general, explanations for machine learning models can be broadly classified into two main types:

1. *Global explanation*, which aims to give a general explanation considering the whole data population
2. *Local explanation*, which focuses on answering specific questions (e.g., "Why a loan was not approved for the customer John?")

The tools available for explanations follow concepts and theories of these two main types. Although there are a plenty of tools available for these two types of explanations, most of the output visualization and verbalization provided are not of easy understanding by non-expert users. To address this issue (i.e., to enable non-expert users to understand the output visualization and verbalization), we present an explainable artificial intelligence solution for providing human-friendly explanations to predictive analytics on big data for both expert and non-expert users.

Machine learning models are commonly classified into two different categories regarding their interpretability:

1. *Crystal-clear models*—such as linear regression and decision trees—which are self-explanatory and do not require the application of XAI techniques to explain them.
2. *Black-box models*—such as random forest and artificial neural network (ANN)— which can be complex to explain themselves. Consequently, they need XAI techniques to explain the results. The complexity of the black-box models makes them achieve higher accuracy when solving complex problems. Thus, XAI may serve to make black-box models more interpretable and avoid the trade-off between accuracy and interpretability.

To evaluate the practicality and usefulness of our XAI solution, we conduct a case study on applying the explanations to a random forest customer churn predictive model. Churn is the rate of customers who stopped using a service or product in a given time frame. The possibility of predicting customer churn can bring a competitive advantage to the business in many different domains such as banking, telecommunications, retail, and education. This kind of strategic knowledge can raise the possibility to prevent and retain potential attrition of customers. Machine learning models have the power to automate the process of identifying those customers, learning from historical data the nuances that differentiate the ones who stopped using a service or product from those who are still loyal. In our particular case study, we used data from customers of a financial institution.

Our solution involves two main components: (i) a back-end component where the machine learning model runs and the explanations' processing occurs and (ii) a front-end component that comprehends the explanation web interface. Hence, in this chapter, *our key contributions* include:

- Creation of a solution that integrates different techniques to facilitate the use and understanding of machine learning reasoning for non-expert users.
- Enhancements in the way explanations are processed and presented for some of the state-of-the-art techniques.

The remainder of this chapter is organized as follows. The next section explains important background concepts for understanding the remaining sections. Section 4.3 describes related works. Section 4.4 presents our explainable AI web interface and its application for a customer churn predictive model. Section 4.5 shows the results of evaluation on our web interface XAI. Finally, conclusions are drawn in Sect. 4.6.

4.2   Background

4.2.1   Shapley Values

Shapley values [46] is a method based on cooperative game theory that calculates the contribution each player had in the final score of a game. In the context of machine learning, each feature is a player of this game and has a contribution to the final prediction. The contribution of each feature is determined based on its average marginal contribution, calculating how the feature affects the result of the prediction when it is present or not for different coalitions of the remaining features [40]. The average marginal contributions are calculated with the following equation:

$$\phi_i(N, v) = \sum_{S \subseteq N \setminus \{i\}} \frac{|S|! \, (|N| - |S| - 1)!}{N!} [v(S \cup \{i\}) - v(S)], \qquad (4.1)$$

where:

- ϕ_i is the average marginal contribution of player i.
- N is the number of players.
- v represents the game.
- S are sets of different coalitions.

To better illustrate Eq. (4.1), we adapted an example that elucidates the Shapley formula and theory for explaining a taxi fare [26]. Imagine that three riders (A, B, and C) share a taxi, and the fare varies according to the distance from the starting point to each home, as illustrated in Fig. 4.1. Consider that the riders pay their portion of the fare right at the beginning, when they first get in the car. The payment order can vary. For instance, (i) A pays, then B and C, or (ii) B then A and

Fig. 4.1 Shapley values illustration for a taxi fare

Table 4.1 Fare share for different coalitions of payment

S	A	B	C
(A, B, C)	$10	$10	$20
(A, C, B)	$10	$0	$30
(B, A, C)	$0	$20	$20
(B, C, A)	$0	$20	$20
(C, A, B)	$0	$0	$40
(C, B, A)	$0	$0	$40

Table 4.2 Average marginal contribution for each rider

	A	B	C
ϕ	$3.33	$8.33	$28.33

C, representing the different coalitions of Eq. (4.1). Table 4.1 depicts what would happen with the fare for each different coalition:

1. For a first coalition, if the riders pay in the order of distance, then A pays $10, B pays $10, and C pays $20 given the total of $40 that comprehends the amount charged by the taxi driver from the start point to the final destination where is C's home.
2. For a second coalition (A, then C and B), A pays $10, C pays $30 but B pays $0.
3. For a third coalition (B, then A and C), B pays $20, A pays $0, and C pays $20.
4. For a fourth coalition (B, then C and A), B pays $20, C pays $20 but A pays $0.
5. For a fifth coalition (C, then A and B), C pays $40, but A pays $0 and B also pays $0.
6. For a sixth and final coalition (C, then B and A), C pays $40, but B pays $0 and A also pays $0.

The average payment for the different coalitions of each rider is what constitutes their marginal contribution. Table 4.2 shows each rider's marginal contribution and the fair amount each one has to pay according to the Shapley values.

As a preview, we use the Shapley values in the current work for global and local explanations. We also adapt the SHapley Additive exPlanations (SHAP) package [37], which applies the theory based on these Shapley values.

4.2.2 Types of Explanation Techniques

There are two key types of explanation techniques:

- *Local explanation* addresses the interpretability of a specific instance of the dataset. The idea is to understand the reasoning that the model applied to a particular instance. This kind of explanation answers questions (e.g., "Why the loan was refused to customer John?").
- *Global explanation* comprehends explanations of the reasoning adopted by the model for most of the patterns learned during the training process. It is useful in cases where explanations in regards to the whole data population are needed (e.g., climate change, economic predictions [54]).

4.3   Related Works

4.3.1   Shapley Additive Explanations

Shapley values have applications to global and local interpretation techniques, which we adapt in this chapter. To elaborate, Lundberg and Lee [37] proposed an extensively used method for interpreting machine learning models with the application of Shapley values theory. Their proposed SHapley Additive exPlanations (SHAP) serve as a unified method of identifying feature importance on the predictions, considering other related works (e.g., LIME [44], DeepLIFT [47]). They compared their SHAP with LIME and DeepLIFT in terms of computational efficiency and how intuitively the explanations were to humans. To measure the consistency with human intuition, they compared the explanations given by the methods with the explanations given by users who understand the data under the experiment. The closer the explanation method has to the humans' explanation, the better the method in terms of accuracy. Results showed that the new approximation method proposed in SHAP uses fewer evaluations to calculate the feature importance, and it has high accuracy. Human subjects' evaluation also showed that SHAP was more intuitive to human understanding than the other previous methods.

Recently, Kumar et al. [28] argued that Shapley values do not provide human-friendly explanations, which are better satisfied with contrastive explanations. Kaur et al. [25] evaluated the level of understanding of interpretability tools by data scientists. Evaluation results revealed that most participants had a wrong interpretation of results and did not use the tools in the way researchers had in mind when they designed them.

In this chapter, we also used the freely available SHAP package. Recall that SHAP was demonstrated to be superior to other approaches in terms of explanation accuracy and computational efficiency [37]. However, Kumar et al. [28] and Kaur et al. [25] observed that SHAP can still be challenging for non-expert users to understand SHAP. This partially motivate our current work on improving SHAP to be more human-friendly. See Sect. 4.4.

4.3.2 Contrastive Explanations

Contrastive or counterfactual explanations refer to the set of techniques that explain the outcome of a specific instance based on what should be done differently for changing the current prediction. It infers the smallest number of changes necessary in the values of the features to modify the prediction outcome. This technique is human-friendly because humans naturally tend to use counter-facts in order to explain facts [40].

Wachter et al. [50] revealed that, for some real-world problems, the counterfactual explanation with the smallest number of changes may not be feasible to turn into action. They ensured that, for these scenarios, a higher number of alternative counterfactuals must be available to make possible selecting the one that better fits the reality. Mothilal et al. [41] extended Wachter et al.'s work and focused on a method of generating a high number of counterfactual explanations that are feasible and respect real-world constraints.

Dhurandhar et al. [14] proposed an approach called contrastive explanations method (CEM) for contrastive explanations in neural networks that has two main components:

- *Pertinent negatives* (*PN*), which highlights the features that are missed in the instance prediction that would change the model outcome.
- *Pertinent positives* (*PP*), which highlights the critical features that contributed to the current outcome

Image and tabular data were inputs for their experiments. Human subjects analyzed explanations for the tabular data outcome, and they evaluated CEM as superior to LIME and Layerwise Relevance Propagation (LRP).

Jia et al. [23] observed that most works in the literature generate the counterfactual instance—also called synthetic neighbour—with the perturbation of some of the features and calculating the distance between the counterfactual instance and the original one. However, given that some synthetic instances were observed to be of "bad" quality that would result in inadequate explanations, they proposed a framework for generating good-quality synthetic neighbours and consequently better model explanations.

van der Waa et al. [49] proposed a method—called *local foil trees*—for finding contrastive explanation using decision trees. The approach has two main components: the fact (the true output class) and the foils (the contrastive class). Considering the case study of churning prediction used in our current work, imagine that we have the class boundary showed in Fig. 4.2(a) and the data point highlighted by the green square is the instance we want to generate the explanations. The first step consists of training a decision tree, using the foil class data points (representing loyal customers). This training method is the one-versus-all approach, in which the decision tree will learn to classify loyal or churn. The data points closer to the point of interest (fact) have a high weight on the decision tree. With the decision tree trained, the selected data point (highlighted by the green square) should be input into

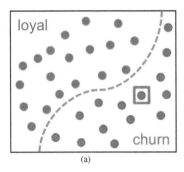 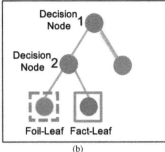

(a)   (b)

Fig. 4.2 Local foil tree method. (**a**) Dataset class boundaries. (**b**) Decision tree: fact leaf, foil leaf, and decision nodes

the trained model and then located in the tree, as shown in Fig. 4.2(b). Then, the foil leaf (as highlighted by a dashed blue square in the figure) with the shortest path from the fact leaf is chosen as the foil. The last step checks the differences between the two selected paths' rules and only accounts for the common features between the two, and all the same parent decision nodes are removed. In our example shown in Fig. 4.2(b), decision node 2 would be excluded, and decision node 1 would be considered to generate the contrastive explanation. The explanation consists of contrasting the values of the fact and foil used on their decision nodes, and the output is in the form "The model predicted A instead of B because feature x is greater than $\langle number \rangle$ and feature y is smaller than $\langle number \rangle$."

4.3.3   XAI User Interfaces

Hohman et al. [21, 22] proposed interactive user interfaces (Gamut, as well as TeleGam) to explain local and global explanations of classification predictions generated by Generalized Additive Models (GAM). The evaluation of the interface was done through human subjects, recruiting machine learning experts, and practitioners. Results showed that the interface improved data professionals' capacity to interpret the model results. Most participants expressed a high interest in having a tool like the one proposed in their daily activities.

Adams and Hagras [2] used a proprietary tool called Temenos to explain the outcome of the predictions of their fuzzy logic model. However, their Temenos appears to be business-specific with a focus on the banking sector.

Wang et al. [52] proposed a theoretical framework that aims to guide model explanations following human reasoning concepts. The framework has four main divisions:

1. Human reasoning and necessity of explanations
2. Ways that people actually reason

3. Ways that XAI generates explanations
4. Ways that XAI supports reasoning

To test their theoretical framework's effectiveness, they developed an XAI explanation dashboard applied to a medical diagnosis model. As they had domain-expert clinicians to test their interface, they had a practical experiment of the proposed theoretical framework. With the user's evaluation, they could see the flaws and improvements needed on the explanations provided. We will refer to this framework as the XAI Diagnostic in the remainder of this chapter.

4.4 Our Explainable AI Web Interface

Recall from Sect. 4.3 that some researches have shown that many existing explainable AI tools may not provide easy of understanding explanations. Moreover, these tools may also require programming knowledge to manipulate the libraries that the techniques were implemented. To address these problems, we present a solution that integrates different explainable approaches in a web interface that creates an abstraction layer between the methods and the non-expert users. We also enhance the generation of explanations for some available state-of-the-art techniques.

As an overview, our explainable artificial intelligence (XAI) system in the current work consists of two main parts:

1. The *back-end component*, which comprehends the background solution's architectural piece, where the processing and generation of the explanations happen
2. The *front-end component*, which brings the interface that the users interact for understanding the model reasoning

4.4.1 Back-End Component

Figure 4.3 shows the overall architecture of our XAI system, in which the back-end component links several pieces to reside the data, preprocess, and run the model, compute explanations, store the results, and have a web framework for the integration of front and back ends. Specifically:

1. The **machine learning model and prediction**, which attest our proposed solution's explanation capabilities. It predicts a financial institution's customer churn as a case study. We build a data pipeline to get the data ready for the model predictions, which involved data exploration, labelling, cleaning, engineering, and data re-sampling. We apply a random forest algorithm because it is a black-box model but led to good performance in various applications. We also measure the model's performance through the recall and precision metrics, which were 80% and 72%, respectively.

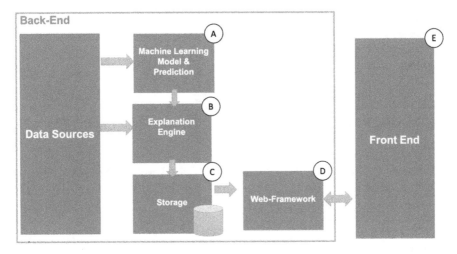

Fig. 4.3   The proposed solution architecture

2. The **explanation engine**, which computes the explanations. The two main processing are:

   - The Shapley values, which generates local and global explanations
   - The contrastive explanations, which aim for the model recommendation

3. The **storage**, which is a database to store the computed explanations. These explanations can then be retrieved by the front-end interface.
4. The **web framework**, which integrates the explanation engine storage results with the front-end interface. The web framework is also responsible for applying any necessary rule on the data before presenting it to the front-end component. One of the well-known paradigms to decouple each one of these parts of the web framework is the Model, View, and Template (MVT), where:

   - *Model* represents the database component.
   - *View* contains logic and actions performed by the server. This layer interacts with the model, applies any necessary logic to the data, and returns the results to the templates.
   - *Template* contains the interfaces.

4.4.2   Front-End Component

The **front-end** component contains the web interface itself where the users can interact and search for interpretations of specific instances or global explanations. As described earlier, this corresponds to the abstraction layer between the users and the techniques. The front-end component provides the following screens:

1. Home and Expected Loss
2. Local Feature Importance
3. Global Feature Importance
4. Model Recommendation

4.4.2.1 Home and Expected Loss

For most business domains, a machine learning model that only classifies observations is not enough. For instance, knowing that a customer will churn or if an employee is going to leave the company is not sufficient for setting a strategic plan of action. There is a need to set strategies and priorities based on the probability of an event to happen and its monetary impact in case it turns into reality (e.g., list Client X with $20M investment and having 80% churn probability before Client Y with $0.2M investment having 90% churn probability). Visual colours to differentiate observations based on the event probability serve to elucidate the appropriate kind of actions.

The Home and Expected Loss screens allow decision-makers to prioritize their actions according to the expected loss, which results from the product of the probability of churn predicted by the model and the monetary value the customer has for the company. Figures 4.4 and 4.5 show the screens applied to our case study, grouping customers according to the risk of churn—which is high (as highlighted by the red colour), medium (as highlighted by the amber colour), or low (as highlighted

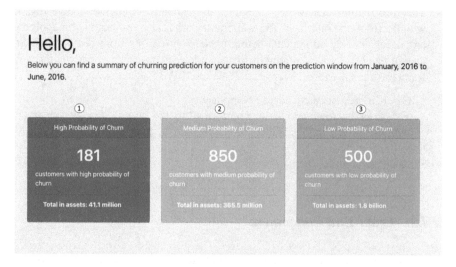

Fig. 4.4 The Home screen. Boxes ①, ②, and ③ with summaries of customers in each group of risk and the total monetary value they represent. These boxes are also clickable leading to the corresponding tab in the Expected Loss screen (as shown in Fig. 4.5)

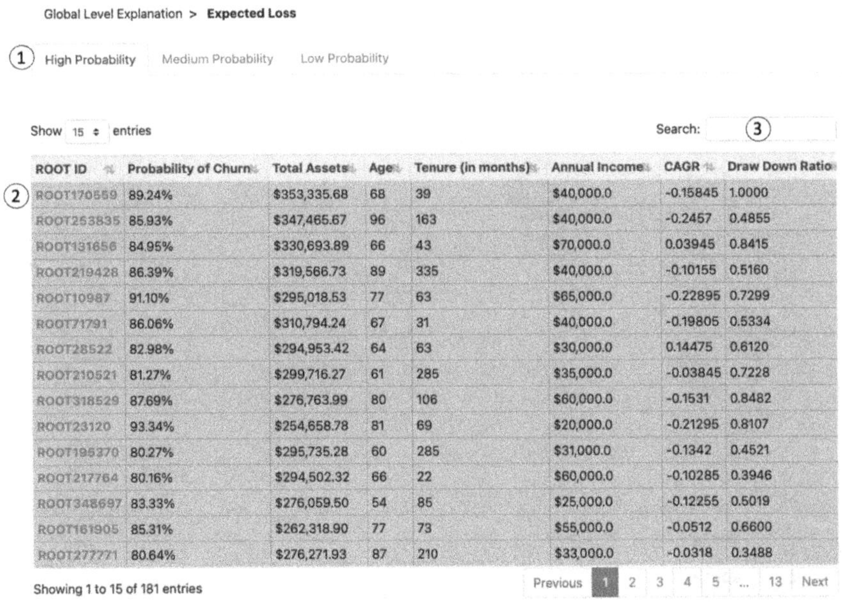

Global Level Explanation > **Expected Loss**

① **High Probability**   Medium Probability   Low Probability

Show 15 ⇕ entries   Search: ③

	ROOT ID ⇅	Probability of Churn ⇅	Total Assets ⇅	Age ⇅	Tenure (in months) ⇅	Annual Income ⇅	CAGR ⇅	Draw Down Ratio ⇅
②	ROOT170559	89.24%	$353,335.68	68	39	$40,000.0	-0.15845	1.0000
	ROOT253835	85.93%	$347,465.87	96	163	$40,000.0	-0.2457	0.4855
	ROOT131656	84.95%	$330,693.89	66	43	$70,000.0	0.03945	0.8415
	ROOT219428	86.39%	$319,566.73	89	335	$40,000.0	-0.10155	0.5160
	ROOT10987	91.10%	$295,018.53	77	63	$65,000.0	-0.22895	0.7299
	ROOT71791	86.06%	$310,794.24	67	31	$40,000.0	-0.19805	0.5334
	ROOT28522	82.98%	$294,953.42	64	63	$30,000.0	0.14475	0.6120
	ROOT210521	81.27%	$299,716.27	61	285	$35,000.0	-0.03845	0.7228
	ROOT318529	87.69%	$276,763.99	80	106	$60,000.0	-0.1531	0.8482
	ROOT23120	93.34%	$254,658.78	81	69	$20,000.0	-0.21295	0.8107
	ROOT195370	80.27%	$295,735.28	60	285	$31,000.0	-0.1342	0.4521
	ROOT217764	80.16%	$294,502.32	66	22	$60,000.0	-0.10285	0.3946
	ROOT348697	83.33%	$276,059.50	54	85	$25,000.0	-0.12255	0.5019
	ROOT161905	85.31%	$262,318.90	77	73	$55,000.0	-0.0512	0.6600
	ROOT277771	80.64%	$276,271.93	87	210	$33,000.0	-0.0318	0.3488

Showing 1 to 15 of 181 entries   Previous **1** 2 3 4 5 ... 13 Next

Fig. 4.5 The Expected Loss screen. ① Tabs to separate the instances according to the group of risk. ② There are links for each of the instances leading to the corresponding explanation in the Local Explanation screen (as shown in Fig. 4.6). ③ The search function enables users to search for specific instances

by the green colour). As a preview, in our case study, the group of customers highlighted in red has 80% or higher probabilities to churn, whereas the group of customers highlighted in amber has between 50% and 79% probabilities to churn. The group of customers highlighted in green has lower than 50% probabilities to churn. These values can be parameterized according to specific needs.

4.4.2.2   Local Feature Importance

Recall from Sect. 4.2 that local explanation techniques aim to explain decisions made for specific instances. As a preview, in our case study, we use this screen to explain why the model predicted a particular customer has high (or medium or low) probabilities of churning. The Local Feature Importance screen brings for each instance the attributes that contributed positively and negatively to the model prediction outcome. We limit the number of features used for explanations because researches have shown that people have better comprehension with explanations that have fewer causes. Thus, we apply a feature selection technique to select the most important features, which are shown in decreasing order of importance in the y-axis of the tornado plot as shown in the top ① of Fig. 4.6. The user can also search for specific instances or switch between them as shown in the bottom table

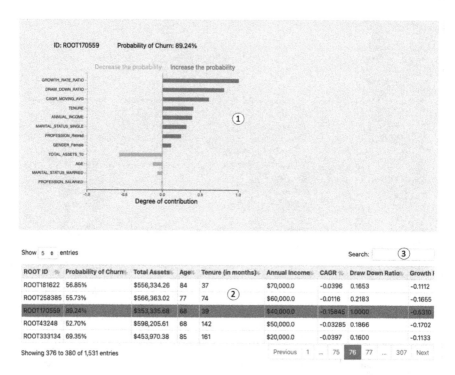

Fig. 4.6 The Local Feature Importance screen. ① The tornado plot displays the positive and negative features contributing to the model outcome. ② The table allows users to change the explanation to another instance and also visualize the feature values for the instances. ③ The search function enables users to search for specific instances

② of the figure. This bottom table facilitates the interaction between the user and the interface, allowing them to access the raw data to understand a customer profile and change between instances easily.

The computation of the Local Feature Importance for each instance in this screen is based on Shapley values. Recall from Sect. 4.3 that the SHAP package computes the importance of the features based on their marginal contribution. When presenting the SHAP value results for end users, they were observed to have a hard time following the meaning of the numbers without understanding the intuition behind the Shapley values theory. Also, the labelling style and the shape of the chart available in the current library bring challenges for them to interpret it. Hence, a more straightforward way to comprehend the numbers and the graph was necessary. First, we normalize the marginal contribution results for the features set for each one of the instances i. The normalization scaled the values in a range between -1 and 1, as shown in Eqs. (4.2) and (4.3):

$$standard_i = \frac{\text{Features List}[i] - \text{Features List.min}()}{\text{Features List.max}() - \text{Features List.min}()} \qquad (4.2)$$

$$scale_i = standard_i \times (maxRange - minRange) + minRange. \quad (4.3)$$

On this scale, -1 means a high negative contribution, and 1 a high positive contribution. Although the Shapley values' calculation results are presented using a distinct approach for enhancing interpretability, each feature's degree of importance in a given explanation is preserved. Moreover, we create a tornado plot with the new range of values in the x-axis, showing the negative or positive contributions the features had in the model prediction outcome. For comparison, Fig. 4.7(a) shows the original plot available in the current library, and Fig. 4.7(b) shows an example of our modified version.

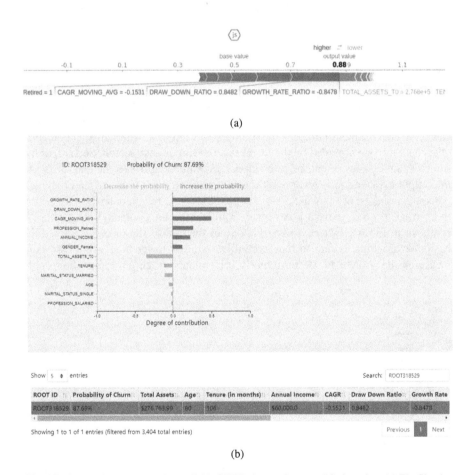

(a)

(b)

Fig. 4.7 Comparison between the available SHAP chart and our modified version. (**a**) The Shapley values chart in SHAP package. (**b**) The modified Shapley values chart

4.4.2.3   Global Feature Importance

The Global Feature Importance screen gives the user a general panorama regarding each feature's relevance for the model when reasoning the classification decisions. This screen gives quick information on the set of features, from the most to the least important, that the model reasoned its decisions. It can also serve as an agile way to identify biased models.

The computation of the importance of the features is also based on the Shapley values theory. The normalization of the values follows the same explanation given in Sect. 4.4.2.2, with the only modification being in the range of the values that in this case ranges from 0 to 1 (where 0 meaning no contribution, and 1 a high contribution). Figure 4.8 shows details of this screen.

4.4.2.4   Model Recommendation

In the Model Recommendation screen, no visualization is used. Instead, verbalization is used for explaining the instances. The information given on this screen can serve two different purposes. First, it can be used as an alternative explanation other than the one provided by the Local Feature Importance screen. Second, as the screen's name states, it can serve as model recommendations to the end users. As the output explanation is given with the necessary changes to modify a prediction outcome, it can be used by users—when possible—as a recommendation for action (e.g., what to do to prevent a customer from leaving the company). The example is shown in Fig. 4.9, where it says that one of the factors the model predicted the customer as churn is due to the "growth rate ratio" lower than 1.37. This ratio measures the growth in the customer's investments. Such an explanation could

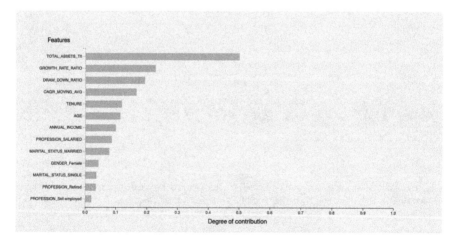

Fig. 4.8 Global Feature Importance screen

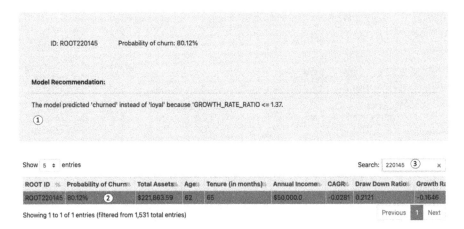

Fig. 4.9 Model Recommendation screen. ① The verbalization explanation contrasting the fact and the foil. ② The table allows users to change the explanation to another instance and also visualize the feature values for the instances. ③ The search function enables users to search for specific instances

trigger the financial advisor to give the customer a call to check if he was satisfied with his investments' growth level or if he wanted to make changes in his portfolio. This kind of proactive action has the potential to prevent a customer from leaving the company.

The recommendations given in this screen are based on contrastive explanation using the local foil trees technique described in Sect. 4.3. Among the different techniques presented in Sect. 4.3, we chose to enhance and integrate the local foil trees method because it is model-agnostic, which means we can apply the technique for a diversity of models. Second, there is no synthetic data points generation, as it uses the contrastive class's actual data points. Lastly, the form of explanations is easy to understand. In our proposed solution, we improved some aspects of the local foil trees method's explanation. We formatted the numerical features' values and adapted the output explanation for categorical variables that were also previously based on numbers. The explanations for categorical features are now in the format "The model predicted X instead of Y because the customer is ⟨is or is not⟩ categorical feature value."

4.5 Evaluation

To evaluate our proposed web interface that integrates and enhances the state-of-the-art machine learning explanation techniques, we compared the number of functionalities available and its differences between our solution and the existing ones.

Table 4.3 Comparison between our proposed solution and the existing ones

Solution	Functionality					Audience
	Global expl.	Local expl.	Contrastive explanation	Search table	Target instance capability	
TeleGam [21]	✓	✓	✓		✓	Experts
Gamut [22]	✓	✓	✓	✓	✓	Experts
XAI Diagnostic [52]		✓	✓		✓	Non-experts
Our proposed solution	✓	✓	✓	✓	✓	Non-experts

As we can observe from Table 4.3, we compare the solutions based on the different types of explanations available, target instances capability, search functionalities, and audience. For instance, Wang et al.'s **XAI Diagnostic** solution [52] is the one that also focused on non-expert users when producing the explanations. However, it does not provide global explanation techniques in the interface. For local explanations, it uses Shapley values to show feature importance (which is the same technique we used in our solution). The model used for explanation is a multi-class model, with five different classes. Unlike our solution, labelling and order based on the features' importance are not presented. Moreover, Wang et al. [52] observed that some users felt that each feature in isolation was responsible for a prediction, which can be attributed to the lack of appropriate labels. Moreover, there are counterfactual explanations in the bottom part of the dashboard, which shows many different scenarios that could be formed with different feature values. In contrast, our solution focused on a technique that shows only one contrasting scenario for each instance, making it not overwhelming to the end users. Furthermore, the XAI Diagnostic solution does not have an easy way of selecting instances based on feature values or filtering capabilities as the interactive table in our solution provides.

Hohman et al.'s **Gamut** [22] offers the same set of functionalities as our solution, but with differences in how the explanations are given and focused on data scientists and machine learning practitioners. Specifically, Gamut gives the global explanation by partial dependence plots (PDP), which are a visual approach to show the interaction between one or more features and the model target outcome. Gamut shows for each feature its interaction with the model outcome. In the same charts used for the Global Explanations, Gamut presents the possibility of creating counterfactual explanations built by their users. For instance, if a user wants to check how a different value for a specific feature would affect the model, he could select the desired value in the chart and check its effect. For the Local Explanation, Gamut uses a waterfall chart that shows how each feature contributed to the specific outcome. The explanations are for a regression model to predict house prices, and they showed how much a feature value added or subtracted to the house predicted price. Although they tested their interface in classification models, but they do not mention how it affects the explanations. Gamut's solution was observed to require more technical expertise because users need to understand PDP plots and hypothesize counterfactuals they want to test. Hohman et al. [22] also observed that some users were initially confused when first started using their Gamut regarding local or global explanations.

In contrast, our solution shows all the features and their level of contribution in a single chart. In our solution, we already present counterfactuals with the set of features and values that would change a prediction. Moreover, we use a tornado plot and show the degree of contribution, either positively or negatively, each feature had for the model outcome. We divide the screens for the specific kind of explanations, making it clear to the users the kinds of explanations they are using. We apply feature selection techniques to show only the most important features as explanations because other studies have shown that explanations based on a high number of causes are not effective for humans [39].

TeleGam [21] is a continuation of the work done in Gamut but with a focus on more verbalization explanations using natural language techniques. The global explanations are now primarily explained by verbal explanations, and PDP plots are shown as an additional resource. The local instance explanation is the same as Gamut. The interactive table is not presented in TeleGam.

From the comparisons, we observe that our solution delivers a satisfying number of functionalities and focuses on explanations for non-expert users. Recall from Sect. 4.3 that we adapt the state-of-the-art techniques to facilitate comprehension by non-expert users, such as normalizing Shapley values to a comprehensible range that do not require an understanding of game theory concepts. We also adapt the visualizations, labelling, and colours. For the contrasting explanations, we adapt the verbalization explanations produced for categorical features and formatted the numerical feature's values. Our interface also serves as an abstraction layer for the non-expert users, who do not need to learn how to code to use the available explanation libraries.

4.6 Conclusions

In this chapter, we presented an explainable artificial intelligence web interface that integrates and enhances the state-of-the-art techniques to produce more understandable and practical explanations to end users. We have shown that the existing methods could be enhanced to facilitate the interpretation of explanations by non-expert users. The results showed that our solution has a satisfying number of functionalities when compared to similar approaches. As ongoing and future work, to further evaluate and demonstrate the effectiveness of our presented solution, we conduct more comprehensive and exhaustive human subject evaluation. Moreover, we incorporate more interactive visualizations to further enhance the easiness of understanding.

Acknowledgments This work is partially supported by (i) the Natural Sciences and Engineering Research Council of Canada (NSERC), (ii) DecisionWorks Consulting Inc., as well as (iii) University of Manitoba. Thanks to G. Barkman, O. Crépeau-Lauzon, S. Delvecchio, W. Kwong, and C. Rieger for their business prospective on customer churn.

References

1. Adadi, A., & Berrada, M. (2018). Peeking inside the black-box: a survey on explainable artificial intelligence (XAI). *IEEE Access, 6*, 52138–52160. https://doi.org/10.1109/ACCESS.2018.2870052

2. Adams, J., & Hagras, H. (2020). A type-2 fuzzy logic approach to explainable AI for regulatory compliance, fair customer outcomes and market stability in the global financial sector. In *FUZZ-IEEE 2020* (pp. 194–201). https://doi.org/10.1109/FUZZ48607.2020.9177542

3. Ahn, S., Couture, S. V., Cuzzocrea, A., Dam, K., Grasso, G. M., Leung, C. K., McCormick, K. L., & Wodi, B. H. (2019). A fuzzy logic based machine learning tool for supporting big data business analytics in complex artificial intelligence environments. In *FUZZ-IEEE 2019* (pp. 1259–1264). https://doi.org/10.1109/FUZZ-IEEE.2019.8858791

4. Al-Mushayt, O. A. (2019). Automating e-government services with artificial intelligence. *IEEE Access, 7*, 146821–146829. https://doi.org/10.1109/ACCESS.2019.2946204

5. Bex, F., Prakken, H., van Engers, T., & Verheij, B. (2017). Introduction to the special issue on artificial intelligence for justice (AI4J). *Artificial Intelligence and Law, 25*, 1–3. https://doi.org/10.1007/s10506-017-9198-5

6. Bortolaso, C., Combettes, S., Gleizes, M., Lartigue, B., Raynal, M., & Rey, S. (2020). SANDFOX project optimizing the relationship between the user interface and artificial intelligence to improve energy management in smart buildings. In *HCII 2020—Late Breaking Papers: Multimodality and Intelligence. LNCS* (Vol. 12424, pp. 417–433). https://doi.org/10.1007/978-3-030-60117-1_31

7. Braun, P., Cuzzocrea, A., Keding, T. D., Leung, C. K., Pazdor, A. G. M., & Sayson, D. (2017). Game data mining: Clustering and visualization of online game data in cyber-physical worlds. *Procedia Computer Science, 112*, 2259–2268. http://doi.org/10.1016/j.procs.2017.08.141

8. Buayai, P., Piewthongngam, K., Leung, C. K. Runapongsa Saikaew, K. (2019). Semi-automatic pig weight estimation using digital image analysis. *Applied Engineering in Agriculture, 35*(4), 521–534. https://doi.org/10.13031/aea.13084

9. Chanda, A. K., Ahmed, C. F., Samiullah, M., & Leung, C. K. (2017). A new framework for mining weighted periodic patterns in time series databases. *ESWA, 79*, 207–224. http://doi.org/10.1016/j.eswa.2017.02.028

10. Chen, Y., Leung, C. K., Shang, S., & Wen, Q. (2020). Temporal data analytics on COVID-19 data with ubiquitous computing. In *IEEE ISPA-BDCloud-SocialCom-SustainCom 2020* (pp. 958-965). https://doi.org/10.1109/ISPA-BDCloud-SocialCom-SustainCom51426.2020.00146

11. Choy, C. M., Co, M. K., Fogel, M. J., Garrioch, C. D., Leung, C. K., & Martchenko, E. (2021). Natural sciences meet social sciences: Census data analytics for detecting home language shifts. In *IMCOM 2021*. https://doi.org/10.1109/IMCOM51814.2021.9377412

12. Confalonieri, R., Coba, L., Wagner, B., & Besold, T. R. (2021). A historical perspective of explainable Artificial Intelligence. *WIREs Data Mining and Knowledge Discovery, 11*(1), 1391:1–1391:21. https://doi.org/10.1002/widm.1391

13. De Guia, J., Devaraj, M., & Leung, C. K. (2019). DeepGx: deep learning using gene expression for cancer classification. In *IEEE/ACM ASONAM 2019* (pp. 913–920). https://doi.org/10.1145/3341161.3343516

14. Dhurandhar, A., Chen, P., Luss, R., Tu, C., Ting, P., Shanmugam, K., & Das, P. (2018). Explanations based on the missing: Towards contrastive explanations with pertinent negatives. In *NeurIPS 2018* (pp. 590–601). https://proceedings.neurips.cc/paper/2018/hash/c5ff2543b53f4cc0ad3819a36752467b-Abstract.html

15. Emmert-Streib, F., Yli-Harja, O., & Dehmer, M. (2020). Explainable artificial intelligence and machine learning: A reality rooted perspective. *WIREs Data Mining and Knowledge Discovery, 10*(6), 1368:1–1368:8. https://doi.org/10.1002/widm.1368

16. Fan, C., Hao, H., Leung, C. K., Sun, L. Sun, Y., & Tran, J. (2018). Social network mining for recommendation of friends based on music interests. In *IEEE/ACM ASONAM 2018* (pp. 833–840). https://doi.org/10.1109/ASONAM.2018.8508262

17. Fariha, A., Ahmed, C. F., Leung, C. K., Abdullah, S. M., & Cao, L. (2013). Mining frequent patterns from human interactions in meetings using directed acyclic graphs. In *PAKDD 2013, Part I*. LNCS (LNAI) (Vol. 7818, pp. 38-49). http://doi.org/10.1007/978-3-642-37453-1_4

18. Hamilton, J. D., Leung, C. K., & Singh, S. P. (2020). Identifying the right person in social networks with double metaphone codes. In *IEEE ISPA-BDCloud-SocialCom-SustainCom 2020* (pp. 794-801). https://doi.org/10.1109/ISPA-BDCloud-SocialCom-SustainCom51426.2020.0012

19. Hammami, Z., Mouchaweh, M. S., Mouelhi, W., & Said, L. B. (2020). Neural networks for online learning of non-stationary data streams: A review and application for smart grids flexibility improvement. *Artificial Intelligence Review, 53*(8), 6111–6154. https://doi.org/10.1007/s10462-020-09844-3

20. Haughton, D., McLaughlin, M., Mentzer, K., & Zhang, C. (2015). Movie analytics. https://doi.org/10.1007/978-3-319-09426-7

21. Hohman, F. (2019). TeleGam: Combining visualization and verbalization for interpretable machine learning. In *IEEE VIS 2019* (pp. 151–155). https://doi.org/10.1109/VISUAL.2019.8933695

22. Hohman, F., Head, A., Caruana, R., DeLine, R., & Drucker, S. M. (2019). Gamut: A design probe to understand how data scientists understand machine learning models. In *CHI 2019* (pp. 579:1–579:13). https://doi.org/10.1145/3290605.3300809

23. Jia, Y., Bailey, J., Ramamohanarao, K., Leckie, C., & Houle, M. E. (2019). Improving the quality of explanations with local embedding perturbations. In *ACM KDD 2019* (pp. 875-884). https://doi.org/10.1145/3292500.3330930

24. Jiang, F., Leung, C. K., & Tanbeer, S. K. (2012). Finding popular friends in social networks. In *CGC 2012* (pp. 501—508). https://doi.org/10.1109/CGC.2012.99

25. Kaur, H., Nori, H., Jenkins, S., Caruana, R., Wallach, H., & Wortman Vaughan, J. (2020). Interpreting interpretability: Understanding data scientists' use of interpretability tools for machine learning. In *CHI 2020* (pp. 92:1–92:14). https://doi.org/10.1145/3313831.3376219

26. Knight, V. (2016). Cooperative games. In *Game theory*. https://vknight.org/Year_3_game_theory_course/Content/Chapter_16_Cooperative_games/

27. Kobusinska, A., Leung, C. K., Hsu, C., Raghavendra, S., & Chang, V. (2018). Emerging trends, issues and challenges in Internet of Things, big data and cloud computing. *FGCS, 87*, 416–419. http://doi.org/10.1016/j.future.2018.05.021

28. Kumar, I. E., Venkatasubramanian, S., Scheidegger, C., & Friedler, S. (2020). Problems with Shapley-value-based explanations as feature importance measures. In *ICML 2020* (pp. 5491–5500). http://proceedings.mlr.press/v119/kumar20e.html

29. Leung, C. K. (2019). Big data analysis and mining. In *Advanced methodologies and technologies in network architecture, mobile computing, and data analytics* (pp. 15–27). http://doi.org/10.4018/978-1-5225-7598-6.ch002

30. Leung, C. K. (2021). Data science for big data applications and services: Data lake management, data analytics and visualization. In *Big Data Analyses, Services, and Smart Data* (pp. 28–44). https://doi.org/10.1007/978-981-15-8731-3_3

31. Leung, C. K. (2014). Uncertain frequent pattern mining. In *Frequent Pattern Mining* (pp. 417–453). http://doi.org/10.1007/978-3-319-07821-2_14

32. Leung, C. K., Braun, P., Hoi, C. S. H., Souza, J., & Cuzzocrea, A. (2019). Urban analytics of big transportation data for supporting smart cities. In *DaWaK 2019*. LNCS (Vol. 11708, pp. 24–33). https://doi.org/10.1007/978-3-030-27520-4_3

33. Leung, C. K., Cuzzocrea, A., Mai, J. J., Deng, D., & Jiang, F. (2019). Personalized DeepInf: Enhanced social influence prediction with deep learning and transfer learning. In *IEEE BigData 2019* (pp. 2871–2880). https://doi.org/10.1109/BigData47090.2019.9005969

34. Leung, C. K., Eckhardt, L. B., Sainbhi, A. S., Tran, C. T. K., Wen, Q., & Lee, W. (2019). A flexible query answering system for movie analytics. In *FQAS 2019*. LNCS (LNAI) (Vol. 11529, pp. 250–261). http://doi.org/10.1007/978-3-030-27629-4_24

35. Leung, C. K., MacKinnon, R. K., & Wang, Y. (2014). A machine learning approach for stock price prediction. In *IDEAS 2014* (pp. 274–277). https://doi.org/10.1145/2628194.2628211

36. Leung, C. K., Wen, Y., & Zheng, H. (2021). Conceptual modeling and smart computing for big transportation data. In *IEEE BigComp 2021* (pp. 260-267). https://doi.org/10.1109/BigComp51126.2021.00055

37. Lundberg, S. M., & Lee, S. I. (2017). A unified approach to interpreting model predictions. In *NIPS 2017* (pp. 4766–4775). http://papers.nips.cc/paper/7062-a-unified-approach-to-interpreting-model-predictions

38. Matzka, S. (2019). Explainable artificial intelligence for predictive maintenance applications. In *AI4I 2020* (pp. 69–74). https://doi.org/10.1109/AI4I49448.2020.00023

39. Miller, T. (2019). Explanation in artificial intelligence: insights from the social sciences. *Artificial Intelligence, 267*, 1–38. https://doi.org/10.1016/j.artint.2018.07.007

40. Molnar, C. (2019). *Interpretable machine learning*. https://christophm.github.io/interpretable-ml-book/

41. Mothilal, R. K., Sharma, A., & Tan, C. (2020). Explaining machine learning classifiers through diverse counterfactual explanations. In *FAT* 2020* (pp. 607–617). https://doi.org/10.1145/3351095.3372850

42. Payrovnaziri, S. N., Chen, Z., Rengifo-Moreno, P., Miller, T., Bian, J., Chen, J. H., Liu, X., & He, Z. (2020). Explainable artificial intelligence models using real-world electronic health record data: A systematic scoping review. *Journal of the American Medical Informatics Association, 27*(7), 1173–1185. https://doi.org/10.1093/jamia/ocaa053

43. Peres, R. S., Jia, X., Lee, J., Sun, K., Colombo, A. W., & Barata, J. (2020). Industrial artificial intelligence in Industry 4.0—systematic review, challenges and outlook. *IEEE Access, 8*, 220121–220139. https://doi.org/10.1109/ACCESS.2020.3042874

44. Ribeiro, M. T., Singh, S., & Guestrin, C. (2016). "Why should I trust you?" Explaining the predictions of any classifier. In *ACM KDD 2016* (pp. 1135–1144). https://doi.org/10.1145/2939672.2939778

45. Shang, S., Chen, Y., Leung, C. K., & Pazdor, A. G. M. (2020). Spatial data science of COVID-19 data. In *IEEE HPCC-SmartCity-DSS 2020* (pp. 1370–1375). https://doi.org/10.1109/HPCC-SmartCity-DSS50907.2020.00177

46. Shapley, L. S. (1953). A value for n-person games. In *Contributions to the theory of games, P-295.*

47. Shrikumar, A., Greenside, P., & Kundaje, A. (2017). Learning important features through propagating activation differences. In *ICML 2017* (pp. 4844-4866). http://proceedings.mlr.press/v70/shrikumar17a.html

48. Souza, J., Leung, C. K., & Cuzzocrea, A. (2020). An innovative big data predictive analytics framework over hybrid big data sources with an application for disease analytics. In *AINA 2020*. AISC (Vol. 1151, pp. 669–680). https://doi.org/10.1007/978-3-030-44041-1_59

49. van der Waa, J., Robeer, M., van Diggelen, J., Brinkhuis, M., & Neerincx, M. (2018). Contrastive explanations with local foil trees. In *ICML 2018 Workshop on WHI* (pp. 41-46). http://arxiv.org/abs/1806.07470

50. Wachter, S., Mittelstadt, B., & Russell, C. (2018). Counterfactual explanations without opening the black box: Automated decisions and the GDPR. *Harvard Journal of Law & Technology, 31*(2), 842–887. https://jolt.law.harvard.edu/assets/articlePDFs/v31/Counterfactual-Explanations-without-Opening-the-Black-Box-Sandra-Wachter-et-al.pdf

51. Wan, J., Yang, J., Wang, Z., & Hua, Q. (2018). Artificial intelligence for cloud-assisted smart factory. *IEEE Access, 6*, 55419–55430. https://doi.org/10.1109/ACCESS.2018.2871724

52. Wang, D., Yang, Q., Abdul, A., Lim, B. Y., & States, U. (2019). Designing theory-driven user-centric explainable AI. In *CHI 2019* (pp. 601:1–601:15). https://doi.org/10.1145/3290605.3300831
53. Weihs, C., Jannach, D., Vatolkin, I., & Rudolph, G. (2017). *Music data analysis*. https://doi.org/10.1201/9781315370996
54. Yang, C., Rangarajan, A., & Ranka, S. (2018). Global model interpretation via recursive partitioning. In *IEEE HPCC/SmartCity/DSS 2018* (pp. 1563–1570). https://doi.org/10.1109/HPCC/SmartCity/DSS.2018.00256

Chapter 5
An Efficient Explainable Artificial Intelligence Model of Automatically Generated Summaries Evaluation: A Use Case of Bridging Cognitive Psychology and Computational Linguistics

Alaidine Ben Ayed, Ismaïl Biskri, and Jean-Guy Meunier

5.1 Introduction

5.1.1 Automatic Text Summarization

Automatic summarization has been adopted for the daily running of affairs. Book abstracts on digital bookstores, show trailers, and headlines on TV broadcasts are samples of summaries we deal with regularly [1, 2]. Automatic summarization has commonly been defined as the process of condensing a piece of media form to a shorter version while preserving key informational elements [3]. The spectrum of its application ranges from texts to audio and video media forms. The particular case of automatic text summarization (ATS) refers to creating a concise, reliable, and fluent abstract from a more extended reference text [4].

Following technological improvements, an enormous volume of textual records is publically accessible [5–8]. This massive amount of the available data calls for automatic text summarization, enabling access to only relevant information. [9] argues that automated summarization has concerns deserving addressing despite having been a target of academic research for more than five decades. Also, it states six main arguments why we need automated text summarization. First, abstracts lessen the time spent on reading a more extended text. They make it possible to consume content efficiently. Second, they facilitate the selection process when searching for a document. Third, automatic summarization can likewise make

A. Ben Ayed (✉) · J.-G. Meunier
UQAM, Montréal, QC, Canada
e-mail: ben_ayed.alaidine@uqam.ca; meunier.jean-guy@uqam.ca

I. Biskri
University of Québec at Trois-Rivières, Trois-Rivières, QC, Canada
e-mail: ismail.biskri@uqtr.ca

© The Author(s), under exclusive license to Springer Nature Switzerland AG 2021
M. Sayed-Mouchaweh (ed.), *Explainable AI Within the Digital Transformation and Cyber Physical Systems*, https://doi.org/10.1007/978-3-030-76409-8_5

the indexing process more effective when dealing with massive textual databases. Fourth, it provides us with less biased summaries than those made by humans. Fifth, automatically generated abstracts carry much personalized information, which can be a valuable supplement to question answering frameworks. Finally, automatic text summarization increases the number of texts that can be processed by commercial abstract services.

Depending on the angle of perception, there are many taxonomies of automatic text summarization [10]. One critical criterion to consider when analyzing ATS approaches is the type of the generated output. The latter can be either an extract or an abstract of a source document. Extractive summarization implies that the original text's most significant segments are excerpted to make the abstract. On the flip side, abstractive summarization uses paraphrasing techniques to present the original format's significant issues logically. It produces the original summaries; that is why it is a more challenging task than extractive summarization. The number of documents to summarize is another criterion that categorizes the summarization process into mono-document and multi-document ATS. When taking language as an angle of view, we can distinguish three variants of ATS: (1) mono-lingual automatic text summarization, when the source input and the final output are in the same language; (2) multi-lingual automatic summarization, when the original text is written in more than one language, thus, the final output would be in the corresponding languages; and (3) cross-lingual automatic text summarization, when the generated summary is not in the same language of the source text. The authors in [11] have pointed out key challenges associated with automatically generated summaries evaluation.

5.1.2   Evaluation Protocols of Automatically Generated Text Summaries

Significant advances in the AGSE research area have been made during the last two decades. Various evaluation protocols have been proposed in this context. Furthermore, many evaluation campaigns have been led since early 1996. SUMMAC (the TIPSTER Text Summarization Evaluation) [12], DUC (Document Understanding Conference) [13], and TAC (Text Analysis Conference) [14] are the most far-reaching ones. Notice that the evaluation process can be carried out in reference to a human-made summary. It can also be conducted without an ideal reference [15].

Recall-Oriented Understudy for Gisting Evaluation (ROUGE) is the standard metric for automatically generated summaries evaluation purposes [16]. It compares the generated output to a set of reference human-produced summaries. There are five main variants of the ROUGE metric:

- ROUGE-N [16]: captures the n-gram overlap between the input and output texts; for instance, ROUGE-1 refers to the overlap of unigrams between system and summary references. ROUGE-2 refers to the overlap of bigrams.

- ROUGE-L [16]: measures the longest matching sequence of words using LCS. LCS does not require consecutive matches since it uses in-sequence matches that reflect sentence-level word order. In this case, there is no need to fix a predefined n-gram length since LCS automatically includes the longest in-sequence common n-grams.
- ROUGE-W [16]: is a bunch of weighted LCS-based statistics that hold serial LCSes.
- ROUGE-S [16]: is a skip-gram co-occurrence metric. It captures any pair of words in a sentence in order, allowing for arbitrary gaps. For instance, skip-bigram measures word pairs' overlap with a maximum of two gaps between sentence tokens.
- ROUGE-SU [16]: is a set of skip-bigram plus unigram-based co-occurrence statistics.

Ramirez-Noriega et al. [17] proposed a new variant of the ROUGE protocol that does not involve human-built model summaries (ASHuR). ASHuR checks whether the most informative sentences of the original text were extracted. Informative sentences are selected based on the frequency of concepts they encode, the presence of cue words, and sentence length. RETENTION is another metric of automatically generated summaries evaluation [18]. It has been used in DUC evaluations [13]. It gives insights on to which extent the extracted summary conveys critical information present in the source text. RESPONSIVENESS has also been used in focus-based summarization tasks of DUC and TAC evaluation campaigns [19]. It uses a 5-point ranking scale, indicating how well the summary satisfied many predefined information criteria. The PYRAMID evaluation is another approach that was built upon the same intuition. It uses SCUs (Summarization Content Units) to compute a set of weighted scores [20]. An automatically generated summary containing units with higher weights would have a high PYRAMID score. An SCU weight for a given text unit is relative to its frequency in the human-made summaries. FRESA is another approach that does not involve human-produced reference summaries [21]. It computes a set of divergences among probability distributions. Another evaluation of text summaries without human references approach was recently proposed by Jonathan et al. [22]. It is based on the linear optimization of content metrics using a genetic algorithm.

Lloret et al. [23] give an overview of challenging issues related to summary evaluation research that remains an effortless task. Notice that all of the previously proposed AGSE models were designed according to a classical natural language processing view that involves computer science, mathematical, and linguistic backgrounds. In this chapter, we present a new cognitive and explainable protocol of AGSE. The proposed approach relies on a reading comprehension model that emerged from cognitive psychology research. In the next section, we review the most critical cognitive psychology models of reading comprehension.

5.1.3   Cognitive Psychology Models for Text Comprehension

Several theories about reading comprehension have been proposed since early 1973 [24]. The proposed models tend to analyze different cognitive processes involved in the reading comprehension activity, including recognizing letters and words and getting their meaning, syntactic parsing of sentences, making predictions and inferences, etc. Below, we make a short review of the most crucial cognitive psychology models of text comprehension.

5.1.3.1   The Resonance Model

The reader's mental presentation is the center of focus of the resonance model. It may be accessible in part as the reading progresses. This intuition is exemplified by the fact that sand propositions may remain in the working memory since they are essential to the text, and those concepts considered secondary are forgotten. The latter ones may be reinstated through reactivation: being instigated by a sentence that is being read. Reinstatement is either top-down or bottom-up. The top-down interpretation precedes the argument that readers try to establish a relationship between incoming text statements and earlier ones. When a connection cannot be established between the working memory and mental representation of a text, it calls for earlier reinstatement of the reader's working memory to ensure a link. The bottom-up interpretation claims that there is nothing like active search processes. Hence, "elements from current sentences activate previous statements when reinstating them to the working memory." The latter assumption has been affirmed by an earlier study, which found out that a reader's mental presentation of a text is prone to resonate with the elements of a sentence being processed [25]. The latter finding led to the establishment of the resonance model [26].

5.1.3.2   The Landscape Model

The landscape model focuses on the construction of a relatively stable memory representation of a text, which is an essential facet of the comprehension process [27]. The landscape model simulates how prominent text items are being activated, stored, and retention strengthened in memory.

5.1.3.3   The Langston and Trabasso Model

It has been argued that statements having a robust causal relation to previous story events are usually read faster [28]. They are further often recalled as well as rated as relevant to the text [29]. It has further been argued that when a story is read, the reader can relate it to causal events. The resonance and landscape models have been found to incorporate causality in their explanations. However, they do not simulate it. The Langston and Trabasso model simulates all of the causality-based effects [30, 31].

5.1.3.4   The Construction–Integration Model

The construction–integration model is one of the most influential reading comprehension theories [32]. It simulates many cognitive processes ranging from recognizing words to constructing a representation of text elements. Kintsch assumes that readers build three different mental representations: (i) a literatim representation of the manuscript, (ii) a semantic one that illustrates the essence of the text, and (iii) a situational representation of the situation to which the manuscript holds. The construction–integration model treats the reading comprehension process as much more than just relationships between explicitly mentioned information printed in the text. It casts light on the inferencing sub-process, which either brings relevant background knowledge into someone's subconscious thoughts or generates new knowledge based on what was read.

5.1.3.5   The Predication Model

The predication model was designed to address issues relating to subjectivity [33]. It employs a distributed representation of words as well as propositions. The main idea is to represent discourse as a network made up of connected nodes. The nodes represent discourse items, and the connections give insights into relations between them. Generally, in a localist model, items and relations are presented separately, while in a distributed representation, there is no clear line between them. Therefore, items are represented as vectors, which determine the relations between them. The predication model relies on Latent Semantic Analysis [34] to automatically acquire an objective vector representation of discourse units.

5.1.3.6   The Gestalt Models

Gestalt's model offers an alternative view of Kintsch's external world knowledge that the world of knowledge is formed by accumulating experiences of event sequences in a microworld. The model has proposed that world knowledge is an amalgamation of experiences in a microworld. The model has been criticized for harboring two issues. One, there is no representation of the order of story events. Two, the processing of a story propositional way requires an equal number of computations, which paves the way for the lack of processing time [35, 36].

5.1.3.7   The Golden and Rumelhart Model

Inferencing is one of the most critical sub-processes involved in reading comprehension. Usually, it involves the reader's general knowledge to activate (retrieve)
not explicitly mentioned information in a text. The abovementioned models differ
in their view toward inferencing. In the construction–integration model, text propositions typically retrieve a set of associated propositions from the reader's world
knowledge net. Then at the integration phase, the propositions that are considered
most appropriate to the text are selected. Here inference is taken as a result of
a search process through the reader's world knowledge. However, it has been
identified that one setback for the construction–integration model is the subjectivity
involved in defining the world knowledge net that is included in the model. To
overcome this problem as well as issues related to the order of story events and the
Gestalt model's processing time, Golden and Rumelhart view the inference as a form
of pattern completion. Even though it seals the loopholes created by Gestalt's model,
it has been criticized for involving a switchback from the distributed representation
to the localist one [37, 38].

5.1.3.8   The Distributed Situation Space Model

Kintsch and Welsch [32], Kintsch and Dijk [39], and Kintsch [40] argued that there
are three levels of text: a surface text level, the textbase level, and the situational
level. Considering texts at situational levels has been criticized as it focuses on
knowledge instead of text. Comprehension of texts calls for concepts or propositions
that originate from the reader's knowledge and not from the text that is being
processed.

Gestalts and Distributed Situation Space Models have similarities in that the
amount of knowledge to be implemented is made manageable by letting stories
take place in a microworld and that situations are represented distributively. Also,
the Distributed Situation Space and the Golden and Rumelhart Models share most
architectural assumptions and their mathematical basis from which it follows how
world knowledge concerning relations between storytime steps is implemented and
how the knowledge is applied to the story representation to result in inferences. It
is also important to note that both models take the issue of situation space very
seriously, and it is from the distributed nature of the space that the Distributed
Situation Space Model gets its name [41].

5.1.3.9   The Structure Building Model

The structure building model focuses on casting light on the involved processes
in the comprehension of various media such as texts and pictures [42]. It divides
the comprehension process into three broad sub-processes: (a) setting a foundation
(base) for the text's mental representations, (b) mapping information onto that base,

and (c) shifting the new structures when dealing with new ideas or when new information is not incongruity with the existing one.

5.1.4 Originality of Our Work

The salient outcome of this research is that:

1. It proposes a cognitive protocol of AGSE since it is built upon a cognitive psychology model of reading comprehension.
2. The proposed AGSE protocol gives insights on to which extent criteria of a good summary are met instead of merely focusing on the N-gram overlaps between the original text and the generated output.

Most of the summary evaluation protocols described in Sect. 5.1.2 were designed according to a pure classical natural language processing view that involves computer science, mathematical, and linguistic backgrounds. This chapter presents a new cognitive and explainable AGSE model. The proposed approach relies on a reading comprehension theory that emerged from cognitive psychology research. Furthermore, classic AGSE protocols only focus on N-gram overlaps between the original text and the generated summary. They do not give any insight on to what extent the criteria of a fair resume are met, namely:

- Retention (coverage): The generated output should cover all the concepts reported in the source document.
- Fidelity: The summary should accurately reflect the author's point of view by focusing on salient concepts conveyed in the original text.
- Coherence: The generated summary should be semantically meaningful.

Since previously proposed AGSE protocols merely focus on the N-gram overlaps between the original text and the generated summary, they only reflect on the retention ratio. They cannot check whether the fidelity criterion is met or not: If a newspaper article reports events related to five concepts and a given automatically generated summary focuses on the three marginal ones. It gets a higher relevancy score than another overview focusing on the two most crucial concepts present in the source text despite it is merely focusing on nonessential events.

This chapter presents a cognitive evaluation protocol of automatically generated text summaries. The proposed approach casts light on to which extent both retention and fidelity are met. The technical and mathematical details of the proposed AGSE protocol are detailed in the next section. The conducted experiments and obtained results are reported in the third section. Conclusion and future work are exposed in the fourth section.

5.2 CATSE: A Cognitive Automatic Text Summarization Evaluation Protocol

5.2.1 The Main Idea

The construction–integration (C-I) model of text comprehension comprises two ordered steps: knowledge construction and knowledge integration. During the construction phase, the C-I model generates a propositional network made up of nodes and connections that encode a crude mental representation of the discourse. The connections are meant to reflect any relationship between the discourse elements (Fig. 5.1). Next, we will consider text sentences as elementary discourse elements.

The mental representation is not refined at this stage. An elaborated propositional network is needed to interpret salient knowledge and infer hidden concepts. In the above example, the elaborated propositional network should reflect, in a condensed way, that the civil war is the dominant concept. It should also reveal the hidden semantic link between sentences (a) and (d). In other words, it should help to infer that "many people are dead in Libya because of civil war." In this way, the mind is stimulated as a network, and the comprehension process refers to activating salient knowledge. This activation begins in the construction phase. The integration process refers to the spread of this activation of salient concepts and marginal ones' deactivation across the network (Fig. 5.2).

The propositional network's set of propositions and their associated connections (semantic links) form what we call the microstructure (Fig. 5.3). According to [39], the CRUD semantic representation of the text being read (the microstructure made

(a): *Libya* is an African country.

(b): Libya is suffering from a civil war.

(c): Civil wars trashed many African countries.

(d): Civil wars caused the *death of many people*.

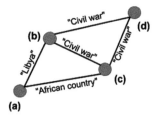

Fig. 5.1 An example of expressing the crude mental representation of the discourse as a propositional network

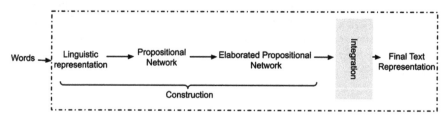

Fig. 5.2 The whole text comprehension process according to the construction–integration model

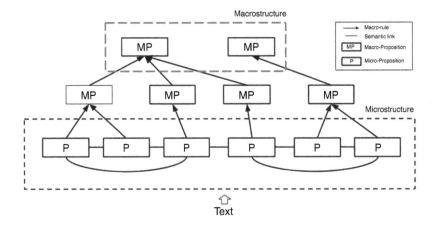

Fig. 5.3 Application of the macro-rules to generate the macrostructure from the microstructure

up of micro-propositions) is too complex to be manipulated and memorized. Thus, the human brain tends to use a set of strategies (macro-rules) that aim to build a more abstracted semantic structure (the elaborated propositional network) containing the text's gist and known as the macrostructure (Fig. 5.3). The macrostructure, made up of macro-propositions, is better suited for memorization or for manipulations that operate on it due to cognitive constraints like the memory size and the complexity of representations.

From a computational perspective, the proposed CATSE protocol generates a feature vector for each input text sentence. Feature vectors encode the different concepts stated by their associated sentences. This representation is better than the propositions-based encoding proposed by the original construction–integration model (described in detail in Sect. 5.2.2.2) as it provides semantic relations between text units. Also, it makes it possible to represent a text unit as a linear combination of concepts. The microstructure is made up of a set of feature vectors encoding text units. Latent Semantic Analysis is used to construct the macrostructure, the compressed form of the microstructure containing salient concepts stored in the semantic memory. Building the macrostructure is a mathematical transformation that consists of mapping feature vectors of the microstructure onto a lower-dimensional space whose unitary vectors encode critical concepts stated in the text. The macrostructure will be stored in the episodic memory.

The integration phase consists of activating text units that encode the salient concepts of the macrostructure. The reader's intention guides this activation process; the CATSE protocol mainly focuses on the retention and fidelity criteria. Coherence is out of the scope of this research. A score assessing the quality of the generated abstract will be produced by the end of the integration phase. It is equal to the average of the activation weights relating to the original text's sentences reported in the generated summary. The activation process will be explained in detail in

Sect. 5.2.3.2. Note that the construction–integration model of text comprehension [32, 40] relies on three different memory structures:

- The semantic memory is simulated by LSA (Latent Semantic Analysis); the semantic similarity between text units is meant to model human associations in semantic memory.
- The working memory audits the mapping between salient concepts and their associated text units.
- The episodic memory keeps track of all concepts or propositions that occur in working memory and their activation values.

5.2.2   Levels of Representation

When we read a text, our mind processes it at three levels:

- The surface structure level
- The intermediate level: the textbase
- The cognitive level: the situation model

Below we describe each level of representation, and we explain how we encode it in our summary evaluation protocol.

5.2.2.1   The Surface Level

The surface structure level is simply the text's words and how they relate to each other at a syntactic level. When we read a sentence, our mind initially understands it at a grammatical level. It checks whether words are in the correct order according to the grammatical rules of a given language. Also, it assesses the level of cohesion in the structure. At this superficial level of representation, our brain tries to understand the information conveyed by sequences of words. If it encounters a text in a language it does not know, it will reject it at the surface structure level. In other words, if we are not familiar with the sequence of terms printed in the text and their syntactic structure, our brain will not waste its cognitive processing time on them. The surface structure level can be considered the first "test" a printed text must go through to be deemed worthwhile for processing. Our CATSE protocol assumes that the summary to evaluate already passed this test since we deal with a by-extraction summary. Furthermore, the goal is to assess the quality of the generated summary, not the original text's quality.

Fig. 5.4 Text unit codification in the construction–integration model

(1) *The fascists have won the elections in El Salvador.*

(2) (i) FASCISTS(x_1)
    (ii) HAVE WON (x_1, x_2)
    (iii) ELECTIONS (x_2)
    (iv) IN (x_2, x_3)
    (v) EL SALVADOR (x_3)

5.2.2.2   The Intermediate Level: The Textbase

The textbase is the second level of information processing in which a proposition codes a basic unit of text. Each proposition refers to a given idea (or concept). Below we present an example of this from [24].

The example (Fig. 5.4) illustrates a sentence (1) and its atomic propositions (2). A complete proposition can replace any variable X_i. Therefore, (ii) could also be written HAVE WON(FASCISTS, ELECTIONS). The theoretical construction–integration model assumes that the whole text is transformed into atomic propositions before the next stage. The CRUD natural language is not suitable for computational processing since it serves many purposes other than the expression of meaning. Kintsch argues that: "propositions are designed to capture those semantic relations that are most salient in text comprehension" [40]. One drawback of the construction–integration model is that it does not technically describe how those propositions would be generated from the CRUD text [40]. Furthermore, when we deal with a long text, we generate many complex propositions, which would explode the processing time and affect the quality of the final output [43].

The central intuition is to break down the text into propositions to get into the essential meaning encoded by each unit (sentence) of the text. Thus, our CATSE protocol encodes each sentence by a feature vector whose coordinates are the *tf-idf* of relevant tokens (words) printed in the text. In this way, each text unit is encoded as a weighted linear combination of cue words (meaning concepts). As a result, instead of breaking down our sentence into n propositions, we encode it by a unique weighted feature vector. Semantic relations between text units can be assessed by computing cosine distance between feature vectors.

5.2.2.3   The Cognitive Level: The Situation Model

The situation model (SM) integrates basic meanings derived from the textbase into our knowledge. Kintsch claims that the SM is continually changing as we read. At any given time, it generally depends on someone's background knowledge and/or the reader's intentions when summarizing a text ranging from achieving given compression ratios to maximizing retention, fidelity, and coherence of the generated summary. In the coming sections, we will suppose that we are working in a closed reasoning word since the goal is to automatically assess the quality of the generated summaries, which are, by construction, supposed only to contain extracted sentences from the text to summarize. They are not supposed to integrate the reader's background knowledge since we are dealing with a "by extraction" summary evaluation protocol. We literally are in an encapsulated knowledge and rote memorization mode; thus, we only limit our focus on the automatic summarizer's intention to generate a fair summary. Modeling the comprehension processes that forge the situation model should be done at the semantic level, especially cognitive modeling. Thus, our CATSE protocol relies on LSA, a powerful model for representing the meaning of words and sentences.

5.2.3 The CATSE Protocol

First, the text to summarize is segmented into units (sentences). Then, a lexicon is built and filtered to discard all universal expressions and terms. Next, the text is coded as an $s \times t$ matrix; s refers to the number of sentences, while t refers to the number of significant unique tokens. We initially map the propositional network (the microstructure) onto a random space during the construction phase. Then, we build the elaborated propositional network (the macrostructure). The latter will be used later to compute a score assessing the generated abstract's quality during the integration phase.

5.2.3.1 The Construction Phase

Each sentence S_i is coded by a sentence feature vector ζ_i of t components. ζ_i components refer to the *tf-idf*s associated to tokens present in a given sentence S_i. The set of ζ_i vectors encodes the micro-propositions, and the feature text matrix, obtained by stacking sentence feature vectors ζ_i as its lines, encodes the microstructure. Afterward, redundant information is coded as ω; the mean sum of sentence feature vectors is ζ_i (Eq. (5.1)). We normalize each sentence feature vector to excrete redundant information since the brain tends not to waste its cognitive processing time (Eq. (5.2)).

$$\omega = \frac{1}{s} \sum_{i=1}^{s} \zeta_i \tag{5.1}$$

$$\varrho_i = \zeta_i - \omega. \tag{5.2}$$

The macrostructure (the new space in which we map the elaborated propositional network) is built by first computing the covariance matrix described in Eq. (5.3). A singular value decomposition will then be performed as described by Eq. (5.4) to construct the macro-propositions (eigenvectors of $\aleph$ associated with the highest eigenvalues).

$$\aleph = \frac{1}{s} \sum_{n=1}^{s} \varrho_n \varrho_n^T = \chi \chi^T \qquad (5.3)$$

$$\chi = \delta.S.\gamma^T \qquad (5.4)$$

$\chi = [\varrho_1, \ldots, \varrho_s]$ in Eq. (5.3). Also, $\aleph$ and χ are, respectively, $t \times t$ and $t \times s$ matrix. Also, dimensions of matrix δ, S, and γ in Eq. (5.4) are, respectively, $t \times t$, $t \times s$, and $s \times s$. Note that, δ and γ are orthogonal ($\delta\delta^T = \delta^T\delta = Id_t$ and $\gamma\gamma^T = \gamma^T\gamma = Id_s$). Additionally:

1. Eigenvectors of $\chi^T\chi$ are columns of γ.
2. Eigenvectors $\chi\chi^T$ are columns of δ.
3. Eigenvalues σ_k of $\chi\chi^T$ and $\chi^T\chi$ are squares of singular values s_k of S.

Eigenvalues σ_k of $\chi\chi^T$ are null when $k > s$ and their associated eigenvectors are unnecessary since $s < t$. So, matrix δ and S can be truncated, and dimensions of δ, S, and γ in (5.4) become, respectively, $t \times s$, $s \times s$, and $s \times s$. Next, the macrostructure Π_K will be built using K eigenvectors δ_i (macro-propositions), belonging to the highest K eigenvalues as shown in Eq. (5.5):

$$\Pi_K = [\delta_1, \delta_2, \ldots, \delta_K]. \qquad (5.5)$$

The construction–integration theory claims that the construction stage refers to (1) building the microstructure (the CRUD semantic representation of the text being read) and (2) transforming it into a macrostructure coding the text's gist. The microstructure and the macrostructure form the textbase, the second level of knowledge processing by our minds. The shifting from the microstructure to the macrostructure is performed by applying a mathematical transformation that maps the original space in which we projected the initial propositional network onto a compressed, more relevant space (the macrostructure) whose unitary vectors are the macro-propositions (constructed vectors that better encode salient concepts yarned by sentences of the text to summarize). The mathematical transformation simulates the macro-rule that aims to build a more abstract semantic structure.

5.2.3.2 The Integration Phase

The integration phase consists of activating text units (sentences) that encode the macrostructure's salient concepts. The reader's intention guides this activation process; the CATSE protocol mainly focuses on the retention and fidelity criteria. Our CATSE protocol's primary concern is to assess the quality of the generated abstract. In other words, it will compute a score that approximates to which extent sentences of the original text are covering all the concepts conveyed in the source text while focusing on the most salient ones. Thus, sentence feature vectors are projected onto the constructed macrostructure and encoded as a linear combination

First macro-proposition	[1 0.09]	[16 0.12]	[3 0.45]	[8 0.48]
Second macro-proposition	[16 0.14]	[3 0.16]	[1 0.21]	[7 0.24]
Third macro-proposition	[4 0.21]	[7 0.32]	[8 0.43]	[3 0.5]
Fourth macro-proposition	[8 0.04]	[1 0.21]	[3 0.24]	[7 0.42]
Fifth macro-proposition	[7 0.5]	[6 0.52]	[4 0.69]	[16 0.92]

[_ _ _] : Computed **Tensor** of the macro-propositions (salient concepts) with a window of 4 text units

[i d] : i is a text unit index, d = cosine distance between a text unit and a macro-proposition (salient concept).

Fig. 5.5 The retention–fidelity tensor

of K macro-propositions as described by Eq. (5.6): the vector $\aleph_{\varrho_i}(k) = \delta_k^T \varrho_i$ provides coordinates of a sentence S_i in the conceptual space.

$$\varrho_i^{proj} = \sum_k \aleph_{\varrho_i}(k)\delta_k. \tag{5.6}$$

Next, the Euclidean distance between a given macro-proposition m and any projected sentence onto the macrostructure is defined and computed as described by Eq. (5.7)

$$d_i(\varrho_m) = \|\varrho_m - \varrho_i^{proj}\|. \tag{5.7}$$

Next, we construct the retention–fidelity tensor (Fig. 5.5) as follows: First, we fix a W window size. W is proportional to the compression ratio. In the below example, W is set to 4. The tensor's first line gives the four text units having the smallest distances to the vector, encoding the first macro-proposition (the most salient concept). The second line shows the same information relative to the second most important concept (macro-proposition). Note here that the order of a given text unit in a given window W depends on its cosine distance to a given macro-proposition. For instance, the first sentence is the best one to encode the first most salient concept, while the eighth sentence is the last one to encode it in a window of four text units. Also, the first macro-proposition is encoded by the eigenvector related to the highest eigenvalue. Thus, it encodes the most salient concept. The second macro-proposition encodes the second most salient concept, and so on. The retention–fidelity tensor simulates the working memory. It will be used later to infer a unified fuzzy retention–fidelity score for each sentence of the source text: First, a retention score is computed to each text unit projected onto the macrostructure. It is defined as the number of times it occurs in the retention–fidelity tensor divided by the number of macro-propositions to be bounded between 0 and 1. The central intuition behind it is that a given sentence having a high retention sore should encode as much as possible of the K macro-propositions of the macrostructure.

$$R_{kw}(s) = \frac{1}{k} \sum_{i=1}^{k} \alpha_i \qquad (5.8)$$

$\alpha_i = 1$ if s occurs in the ith window. If not, it is equal to zero.

Next, we compute a fidelity score, defined as shown in the ninth equation, as the averaged sum of summary sentences' retention coefficients. The fidelity score's central intuition is that text units whose fidelity score is high should encode the most salient concepts stated in the source text. So, they should have minimum distances from the macro-propositions in Eq. (5.7). In other words, the fidelity score gives insights on to which extent a given sentence encodes concepts present in the original text (the macro-propositions) while taking into consideration the salience degree of each one of them. Mathematically, the fidelity score is defined as follows:

$$F_{kw}(s) = \frac{1}{k} \sum_{i=1}^{k} \alpha_i [1 + \frac{1 - \psi_i}{w}] \qquad (5.9)$$

$\alpha_i = 1$ if s occurs in the ith window of the retention–fidelity tensor. If not, it is equal to zero. ψ_i is the rank of s in the ith window.

Next, we use fuzzy logic to compute a unified retention–fidelity score (R-F) for each sentence of the source text. Text units having the highest (R-F) scores will be activated by the end of the integration phase. They will remain in the episodic memory, and they will present candidate sentences of an ideal summary. We opt for the fuzzy logic to compute the unified retention–fidelity score because the brain is a "fuzzy machine." Linguistic variables are input and output variables in simple words (Fig. 5.6). "Low," "medium," and "high" are the linguistic terms used to model the retention and fidelity scores. Afterward, we build a set of rules into the knowledge base in the form of IF-THEN-ELSE structures:

- **Rule 1:** *If the retention score is high or the fidelity score is high, then the R-F score is high.*
- **Rule 2:** *If the fidelity score is medium, then the R-F score is medium.*
- **Rule 3:** *If the retention score is low and the fidelity score is also low, then the R-F score is low.*

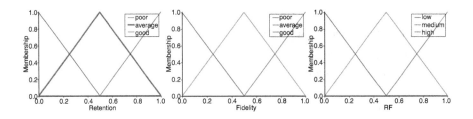

Fig. 5.6 Defining linguistic variables and membership functions

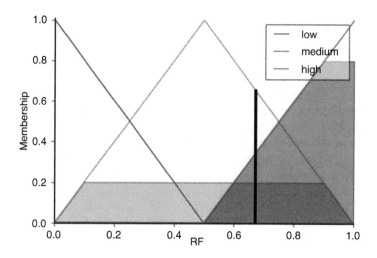

Fig. 5.7 Fuzzy *R-F* score

Next, fuzzy set operations evaluate the previously defined rules to infer the fuzzy values of *R-F* sores. Here, the used operations for "OR" and "AND" are "Max" and "Min," respectively. Afterward, we combine all evaluation results to form final fuzzy *R-F* scores (Fig. 5.7).

Defuzzification is performed according to the membership function for output variables, as shown in Fig. 5.7. A unified retention–fidelity (*R-F*) score is computed for every sentence of the source text. Text units having the highest *R-F* scores will be activated (integrated into the situation model) and stored in the episodic memory. Sentences with low *R-F* scores will be deactivated and forgotten (Fig. 5.8). The CATSE score is equal to the averaged sum of retention–fidelity (*R-F*) scores of sentences chosen to be included in the actual summary.

5.3 Experiments and Results

5.3.1 Datasets

In this chapter, we mainly used three datasets:

- The Timeline17 dataset [44]: It consists of 17 manually created timelines and their associated news articles. They mainly belong to 9 broad topics:

    - BP Oil Spill
    - Michael Jackson Death (M-J)
    - Haiti Earthquake (H-E)
    - H1N1 (Influenza)

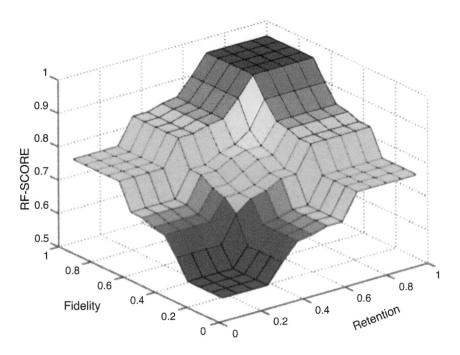

Fig. 5.8 Defuzzification and activation of text units (sentences) having the highest *R-F* scores

- – Financial Crisis (F-C)
- – Syrian Crisis (S-C)
- – Libyan War (L-W)
- – Iraq War (I-W)
- – Egyptian Protest (E-P)

    Original texts belong to news agencies, such as BBC, Guardian, CNN, Fox News, and NBC News.
- • Crisis dataset [45]: It consists of 20463 news articles dealing with the crisis in Egypt, Libya, Yemen, and Syria. Famous news agencies produced the original texts.
- • The EASC dataset [45]: Contains 153 Arabic articles and 765 human-generated extractive summaries of those articles.

5.3.2 Experimental Results

To evaluate the proposed protocol of CATSE, we computed the Spearman correlation between ROUGE and CATSE scores using the three abovementioned datasets. Note that the Spearman correlation of two variables equals the Pearson correlation between their associated rank values: In contrast to the Pearson correlation that assesses linear relationships, Spearman correlation assesses monotonic relationships (whether linear or not). A perfect Spearman correlation of +1 or −1 occurs when each variable is a perfect monotone function of the other. Those correlations were computed using reference texts and automatically generated ones using three summarizers:

- The Luhn summarizer [46]: Luhn summarizer is a naive approach based on tf-idf. It extracts "salient" sentences of the source text. A given sentence's saliency depends on a bunch of meta-heuristics, including cue terms, sentence position in the text, and many other indicators.
- The TextRank summarizer [47]: TextRank builds an undirected graph using text units as vertices. The degree of semantic or lexical similarity between text units is attributed as a weight to vertices edges. The constructed graph is used to build a stochastic matrix. Next, the ranking over vertices is obtained by finding the eigenvector corresponding to the eigenvalue that gives a stationary distribution of the random walk on the graph.
- The LexRank summarizer [48]: It is also a graph-based summarizer. Measured semantic similarities or content overlaps between sentences are defined as weights to graph edges. LexRank uses the cosine similarity of tf-idf vectors in contrast to the TextRank approach that uses a very comparable weight based on the number of shared words between two sentences (generally normalized by the sentences' lengths).

The obtained results are reported in Tables 5.1, 5.2, and 5.3. Notice that if the Spearman correlation score is positive, it means that the two studied variables are positively correlated; if the first variable increases, the second increases. If the first variable decreases, it will be the same scenario for the second one. In contrast, if the Spearman correlation score is negative, it means that the two variables are negatively correlated. Also, if the Spearman correlation score is superior to 0.5, it means that the two variables are highly positively correlated. The obtained results from the 24 experiments led on the eight subsets of the Timeline17 dataset show that the ROUGE and CATSE scores are highly positively correlated in 20 scenarios. They are moderately positively correlated in three scenarios and negatively correlated in

Table 5.1 Spearman correlation between CATSE and ROUGE scores in the Timeline17 dataset

Sub-dataset	Support	Luhn	TextRank	LexRank
BP-Oil	1415	0.74	0.66	0.57
E-P	563	0.81	0.62	0.64
H1N1	215	0.31	0.59	0.61
H-E	125	0.52	0.73	0.88
I-W	125	0.76	0.82	0.81
L-W	325	−0.16	0.23	0.19
S-C	214	0.82	0.86	0.91
M-J	124	0.79	0.83	0.86

Table 5.2 Spearman correlation between CATSE and ROUGE scores in the Crisis dataset

Sub-dataset	Support	Luhn	TextRank	LexRank
Egypt	5110	0.72	0.66	0.64
Libya	5665	0.83	0.79	0.83
Syria	5355	0.51	0.81	0.82
Yemen	4333	0.74	0.69	0.39

Table 5.3 Spearman correlation between CATSE and ROUGE scores in the EASC dataset

Sub-dataset	Luhn	TextRank	LexRank
Art and music	0.11	0.42	0.16
Education	0.31	0.56	0.33
Environment	0.33	0.51	0.52
Finance	0.29	0.61	0.74
Health	−0.21	0.02	0013
Politics	0.16	0.29	0.24
Science and technology	0.52	0.49	0.32
Tourism	0.39	0.58	−0.25
Politics	0.53	0.50	−0.11
Sports	0.34	−0.21	0.41

only in one scenario. The obtained Spearman correlation results using all subsets of the second dataset show that the ROUGE and CATSE scores are highly positively correlated. Better results are obtained when using long texts with many concepts. Our approach assumes that the source text to summarize reports at least two concepts. Thus, it is more suitable for long texts. Experiments conducted on the third dataset show that our method needs more parameter tuning to obtain a realistic assessment of the quality of generated summaries in Semitic languages.

5.4   Conclusion

This chapter focused on a sub-task of ATS, namely automatically generated summaries evaluation (AGSE). We proposed a cognitive and explainable approach of AGSE. The proposed model relies on the Kintsch theory of reading comprehension. Our evaluation protocol is tested and compared to Recall-Oriented Understudy for Gisting Evaluation (ROUGE): a standard approach used to evaluate automatically generated summaries. Conducted experiments on the Timeline17, Crisis, and EASC datasets show that our approach's scores are generally highly positively correlated to the state-of-the-art ROUGE ones. Best results are obtained when using long texts. A refined parameter tuning is needed when assessing the quality of generated summaries in Semitic languages. This research's salient outcome is that it proposes a cognitive and explainable model of AGSE, and it pushes toward cognitive natural language processing. It also demonstrates how cognitive psychology can be used for an explainable artificial intelligence (XAI) approach to justify an AGSE model's scores.

The current version of the proposed evaluation protocol only works with extractive summarization tasks. Now, we are implementing the abstractive oriented version of it. The main idea is the following: after identifying important sentences in the text, the next variant of our CATSE approach will tend to detect the reader's strategy to build the summary, which is viewed as applying adequate macro-rules. Macro-rules are the core of the cognitive processes involved in the summarization activity. [39]. Note that the construction–integration model states three types of macro-rules:

- Deletion: Each text unit containing minor redundant or unrelated details may not be considered part of the summary.
- Generalization: A generic segment may substitute a bunch of text units.
- Construction: Salient text units from retention and fidelity perspectives are stored in the episodic memory and activated later on to be part of an ideal summary.

The CATSE protocol should infer that a particular sentence in the text has been deleted given a text and its summary. In the same way, it should infer whether a sentence is a generalization, and so on. The selection (construction) macro-rule is already simulated using LSA: Each sentence of the summary is semantically compared with each source text sentence. A sentence of the source text will be considered deleted for the deletion macro-rule if no generated output sentence is sufficiently close to it. Similarly, a generalized sentence is a summary sentence close enough to more than one source text sentence.

Acknowledgments The authors would like to thank the Natural Sciences and Engineering Research Council of Canada (NSERC) as well as the Canadian Social Sciences and Humanities Research Council (SSHRC) for financing this work.

References

1. Widyassari, A. P., Affandy, A., Noersasongko, E., Fanani, A. Z., Syukur, A., & Basuki, R. S. (2019). Literature review of automatic text summarization: Research trend, dataset and method. In *International Conference on Information and Communications Technology (ICOIACT)* (pp. 491–496).
2. Wafaa, S. E, Cherif, R. S, Ahmed, A. R., & Hoda, K. M. (2021). Automatic text summarization: A comprehensive survey. In *Expert systems with applications* (Vol. 165).
3. Radev, D. R., Hovy, E., & McKeown, K. (2002). Introduction to the special issue on summarization. *Computational Linguistics, 28*(4), 399–408.
4. Mahak, G., & Vishal, G. (2017). Recent automatic text summarization techniques: A survey. *The Artificial Intelligence Review, 47*(1), 1–66.
5. Liang, H., & Fu, K-W. (2017). Information overload, similarity, and redundancy: Unsubscribing information sources on twitter. *Journal of Computer-Mediated Communication, 22*(1), 1–17.

6. Lee, S. K., Lindsey, N. J., & Kim, K. S. (2017). The effects of news consumption via social media and news information overload on perceptions of journalistic norms and practices. *Computers in Human Behavior, 75*, 254–263.

7. Roetzel, P. G. (2019). Information overload in the information age: A review of the literature from business administration, business psychology, and related disciplines with a bibliometric approach and framework development. *Business Research, 12*, 479–522.

8. Schmitt, J. B., Debbelt, C. A., Schneider, & F. M. (2018). Too much information? Predictors of information overload in the context of online news exposure. *Information, Communication & Society, 21*(8), 1151–1167.

9. Juan-Manuel, T. M. (2014). *Automatic text summarization*. Wiley Publishing.

10. Mani, I. (2001). *Automatic summarization*. John Benjamins Publishing.

11. Saggion, H., & Poibeau, T. (2013). Automatic text summarization: Past, present and future. In T. Poibeau, H. Saggion, J. Piskorski & R. Yangarber (Eds.), *Multi-source, multilingual information extraction and summarization, theory and applications of natural language processing* (pp. 3–21). Springer.

12. Mani, I., Klein, G., House, D., Hirschman, L., Firmin, T., & Sundheim, B. (2002). SUMMAC: A text summarization evaluation. *Natural Language Engineering, 8*(1), 43–68.

13. Over, P. D, Dang, H. T, & Harman, D. K. (2007). DUC in context. *IPM, 43*(6), 1506–1520.

14. *Proceedings of the Text Analysis Conference (TAC)*. NIST (pp. 17–19). Gaithersburg, Maryland, USA (2008)

15. Chin-Yew, L. (2003). Automatic evaluation of summaries using N-gram co-occurrence statistics. In *Proceedings of Language Technology Conference. HLT-NAACL*.

16. Chin-Yew, L. (2004). ROUGE: A package for automatic evaluation of summaries. In *Text Summarization Branches Out* (pp. 74–81).

17. Ramirez-Noriega, A., Juarez-Ramirez, R., Jimenez, S., & Inzunza, S. (2018). ASHuR: Evaluation of the relation summary-content without human reference using ROUGE. *Computing and Informatics, 37*, 509–532.

18. Eduard, H. (2005). Text summarization. In *The Oxford Handbook of Computational Linguistics*.

19. Conroy, J. M., & Dang, H. T. (2008). Mind the gap: Dangers of divorcing evaluations of summary content from linguistic quality. In *Proceedings of the 22nd International Conference on Computational Linguistics, 1*(8), 145–152.

20. Nenkova, A., & Passonneau, R. J. (2004). Evaluating content selection in summarization: The pyramid method. In *HLT-NAACL* (pp. 145–152).

21. Juan Manuel, T. M., Horacio, S., Iria, D. C., Eric, S. J., & Patricia, V. M. (2010). Summary evaluation with and without references. *Polibits, 42*.

22. Jonathan, R. S., Yulia, L., & René-Arnulfo, G. H. (2021). Evaluation of text summaries without human references based on the linear optimization of content metrics using a genetic algorithm. In *Expert systems with applications* (Vol. 167).

23. Lloret, E., Plaza, L., & Aker, A. (2018). The challenging task of summary evaluation: An overview. In *Language Resources and Evaluation, 52*(1), 101–148.

24. Dijk, V., & Kintsch, W. (1983). *Strategies of discourse comprehension*. New York: Academic Press.

25. Albrecht, J. E., & Myers, J. L. (1995). Role of context in accessing distant information during reading. *Journal of Experimental Psychology: Learning, Memory, and Cognition, 21*, 1459–1468.

26. Myers, J. L., & O'Brien, E. J. (1998). Accessing the discourse representation during reading. *Discourse Processes, 26*, 131–157.

27. Van den Broek, P., Risden, K., Fletcher, C. R., & Thurlow, R. (1996). A "landscape" view of reading: Fluctuating patterns of activation and the construction of a stable memory representation. In B. K. Britton & A. C. Graesser (Eds.), *Models of understanding text* (pp. 165–187). Mahwah, NJ: Erlbaum.

28. Myers, J. L., Shinjo, M., & Duffy, S. A. (1987). Degree of causal relatedness and memory. *Journal of Memory and Language, 26*, 453–465.

29. Trabasso, T., & Sperry, L. L. (1985). Causal relatedness and importance of story events. *Journal of Memory and Language, 24*, 595–611.
30. Langston, M. C., & Trabasso, T. (1999). Modeling causal integration and availability of information during comprehension of narrative texts. In H. van Oostendorp & S. R. Goldman (Eds.), *The construction of mental representations during reading* (pp. 29–69). Mahwah, NJ: Erlbaum.
31. Langston, M. C., Trabasso, T., & Magliano, J. P. (1999). A connectionist model of narrative comprehension. In A. Ram & K. Moorman (Eds.), *Understanding language understanding: Computational models of reading* (pp. 181–22). Cambridge, MA: MIT Press.
32. Kintsch, W., & Welsch, D. M. (1991). The construction-integration model: A framework for studying memory for text. In W. E. Hockley & S. Lewandowsky (Eds.), *Relating theory and data: Essays on human memory in honor of Bennet B. Murdock* (pp. 367–385).
33. Kintsch, W. (2001). Predication. *Cognitive Science, 25*, 173–202.
34. Thomas, K., Peter, W., & Laham, D. (1998). Introduction to latent semantic analysis. *Discourse Processes, 25*, 259–284.
35. St-John, M.F. (1992). The Story Gestalt: A model of knowledge-intensive processes in text comprehension. *Cognitive Science, 16*, 271–306.
36. St-John, M. F., & Mc-Clelland, J. L. (1992). Parallel constraint satisfaction as a comprehension mechanism. In R. G. Reilly & N. E. Sharkey (Eds.), *Connectionist approaches to natural language processing* (pp. 97–136).
37. Golden, R. M., & Rumelhart, D. E. (1993). A parallel distributed processing model of story comprehension and recall. *Discourse Processes, 16*, 203–237.
38. Golden, R. M., Rumelhart, D. E., Strickland, J., & Ting, A. (1994). Markov random fields for text comprehension. In D.S. Levine & M. Aparicio (Eds.), *Neural networks for knowledge representation and inference* (pp. 283–309).
39. Kintsch W., & Dijk, V. (1978). Toward a model of text comprehension and production. *Psychological Review, 85*, 363–394.
40. Kintsch, W. (1988). The role of knowledge in discourse comprehension: A construction-integration model. *Psychological Review, 95*(2), 163–182.
41. Frank, S. L., Koppen, M., Noordman, L. G. M., & Vonk, W. (2003). Modeling knowledge-based inferences in story comprehension. *Cognitive Science, 27*, 875–910.
42. Morton-Ann, G. (1995). The structure building framework: What it is, what it might also be, and why. In B. K. Britton & A. C. Graesser (Eds.), *Models of text understanding* (pp. 289–311).
43. Grabowski, J. (1992). Expository text and propositional text processing. In: B. Hout-Wolters & W. Schnotz(Eds.), *Text comprehension and learning from text* (pp. 19–33). Amsterdam: Swets and Zeitlinger.
44. Tran, G. B., Tran, T. A., Tran, N. K., Alrifai, M., & Kanhabua, N. (2013). Leverage learning to rank in an optimization framework for timeline summarization. In *TAIA workshop SIGIR*.
45. El-Haj, M., Kruschwitz, U., & Fox, C. (2017). Using Mechanical Turk to create a corpus of Arabic summaries.
46. Luhn, H-P. (1958). The automatic creation of literature abstracts. *IBM Journal, 2*(2), 159–165.
47. Mihalcea, R., & Tarau, P. (2004). TextRank: Bringing order into texts. In *Proceedings of Empirical Methods for Natural Language Processing* (pp. 404–411).
48. Erkan, G., & Radev, D. R. (2004). LexRank: Graph-based lexical centrality as salience in text summarization. *Journal of Artificial Intelligence Research, 22*, 457–479.

Chapter 6
On the Transparent Predictive Models for Ecological Momentary Assessment Data

Kirill I. Tumanov and Gerasimos Spanakis

6.1 Introduction

At present, user activity and behaviour are being tracked in a number of contexts. Including, for instance, tracking of smartphone usage, mobility, and communication [1], as well as food intake and activity type [2]. Behaviour monitoring is also performed for medical purposes using sensors, for example, to detect development of Parkinson or Alzheimer disease [3]. The common goal of behaviour tracking is identification of the drivers behind the observed activity. Once the drivers are identified, the behaviour can serve, for instance, to predict a person's life outcomes based on personality traits inferred from the behaviour [1] or to provide recommendations to caregivers based on behavioural patterns [3].

Obesity is one of the challenges the modern society is facing [4]. Among other factors, obesity is linked to the: (1) cheapness and availability of the energy-dense food high in sugar, fat, and salt; and (2) clever marketing strategies used by the food manufacturers [4]. Although it is not in the hands of an individual to rule out these two factors, the individual can be supported in adapting own behaviour to cope with them. The adaptation can be achieved by making own informed choices related to food intake and a desired level of physical activity.

However, the process of informing the user about a need to take action to change their behaviour is still not optimal. Once the user behaviour is analysed by an algorithm and a model of the behaviour is constructed, the model itself remains invisible to the user, who only obtains an outcome from the model. In the case of eating behaviour, which is covered in this work, the outcome is an indication whether the user is expected to consume something which is considered

K. I. Tumanov (✉) · G. Spanakis
Department of Data Science and Knowledge Engineering, Maastricht University, Maastricht, The Netherlands
e-mail: k.tumanov@maastrichtuniversity.nl; jerry.spanakis@maastrichtuniversity.nl

© The Author(s), under exclusive license to Springer Nature Switzerland AG 2021
M. Sayed-Mouchaweh (ed.), *Explainable AI Within the Digital Transformation and Cyber Physical Systems*, https://doi.org/10.1007/978-3-030-76409-8_6

an unhealthy option. In this case, the user does not obtain an insight of the context in which the decision is made. This means that the inference process incorporated in the behaviour model does not trigger an equivalent inference response on the user's side.

In this work, we analyse how the behavioural data, obtained via Ecological Momentary Assessment (EMA) can be used to build transparent models for prediction of user behaviour type. We argue that the use of the transparent models can help extending the feedback provided to the user based on the collected EMA data. Below, the relevant concepts are introduced and a brief overview of prior work is provided.

6.1.1 Ecological Momentary Assessment (EMA)

A study of human behaviour requires observation. Normally, the observation cannot be continuously performed by a specialist. Thus, often a person is asked to "log" their own behaviour over a specified period of time on their own. The simplest form of logging is keeping a diary [5] in which the behavioural characteristics of interest are captured. However, the use of simple diaries possesses problems related to the data validity [5]. Namely, the diaries require a person to: (1) remember own behaviour characteristics, (2) accurately fill them retroactively, and (3) fill them regularly [5].

A concept of EMA was introduced by Stone and Shiffman in 1994 to refer to "monitoring or sampling strategies to assess phenomena at the moment they occur in natural settings" [6]. Performing a repeated sampling of behaviour in a typical environment of each participant was hypothesized to improve validity of the collected data and to reduce the participant's need to remember and recall their activity [6]. Therefore, the use of EMA is aimed to avoid the problems linked to the traditional logging with diaries and questionnaires [5].

Rapid development of the digital tools and mobile devices allows automation of the EMA procedures. Meaning that the data can be collected with a smartphone [2, 7]. The use of electronic devices allows automatically reminding a person to report own behaviour without involving a human experimenter [2, 7]. Moreover, the use of a personal mobile device provides a potential to perform covert (or passive) data collection about the person's behaviour [1]. Therefore, management of EMA studies becomes easier and human factors during data entry are reduced.

The EMA data can be successfully used not only for monitoring per se. The data can be used to train models of human behaviour for the use in an Ecological Momentary Intervention (EMI) [8]. During EMI, a person receives real-time feedback based on the analysis of the behaviour data collected until now. The EMI-based Cognitive Behavioural Therapy (CBT) was successfully implemented to change the eating behaviour of overweight adults [7, 9]. It is also imperative for the success of any EMI method that the feedback provided to users is transparent.

6.1.2   Classification of EMA Data

Multilevel modelling [10], logistic regression [11], decision trees [10–13], random forest and bagged boosted trees [11, 14], Hidden Markov Models (HMMs) [15] and Recurrent Neural Networks (RNNs) [16] were previously used for classification of EMA data. The choice of a particular method is largely dependant on the use case and the type of data being acquired through the EMA. Nevertheless, the importance of the ability to explain the classification model was brought up by several authors [10, 13, 15].

EMA data is typically classified post hoc in a Cross-Validation (CV) setting with a sample hold-out strategy. This approach to classification is determined by a common need to test a hypothesis of whether some construct (e.g. depression as in [11]) can be predicted based on the collected data. However, this approach lacks flexibility as: (1) the results can only be obtained at the end of the data collection, (2) all data of a participant is assumed to be available and static, and (3) limited argumentation can be made about the case when new participants are added (which would require testing a participant hold-out strategy). A more flexible alternative, capable of addressing the mentioned limitations, is training/testing the prediction models incrementally, as described in this work.

6.1.3   Model Transparency

Providing a precise and relevant definition of *transparent* algorithms continues to puzzle researchers [17]. Nevertheless, transparency is acknowledged to be one of the key elements constituting user trust in Artificial Intelligence (AI) systems [18]. Availability of the source code and ensuring that the algorithm is following a set of regulatory and societal norms are suggested to provide transparency [17]. However, from the user perspective, these factors are not helpful, and an alternative is to ensure transparency without the source code [17]. Can the predictive models produced by an algorithm be made transparent instead?

One approach to making the models transparent is the use of the network structure to explain relations between variables or user states. Such an approach, for instance, gained popularity in analysis of psychological conditions [19]. The idea is to apply a selected mathematical operator (e.g. a linear correlation) to data to obtain the relations between the variables and then to use network analysis concepts (centrality, cardinality, betweenness, etc.) to characterize the obtained relations. This approach helps visualize complex structures and allows tuning the network analysis parameters for making the phenomena of interest more prominent.

Another approach is to uncover the model decision-making structure to the user. This approach is widely used in process mining; where, first, the data is used to discover a model of the process, and then, the model is used for data conformance checking [20]. Another variant of this approach is presenting the rules uncovered by an algorithm. For instance, a decision tree can be presented to a participant, such that they would be able to see what leads to a particular behaviour. This way a participant can acquire a data-driven, tailored, and compact representation of the model describing their behaviour.

The motivation behind this work was to provide a comprehensive outlook over the use of EMA data to train prediction models which are interpretable, comparable, and plottable. Thus, in this work it is demonstrated how the well-established concepts of model learning, model-based classification, model analysis, and network analysis are used in conjunction with the EMA data to provide a participant with insights of their behaviour.

On the other hand, this work does not aim at providing a detailed analysis of the individual behaviours of participants captured in the used dataset, since this was done previously in [12, 13, 21–23]. However, the only classification results reported previously for the used EMA dataset were obtained in a CV setting without a distinction across participants [14]. Thus, in this work the initial performance values are reported for the case of personalized models used in an incremental training/prediction setting with a cold-start and no strong emphasis was put at optimizing the prediction performance of the used classification algorithms.

The remainder of the work is structured as follows. In Sect. 6.2, an EMA dataset used throughout this work is introduced. In Sect. 6.3, a description of the classification algorithms used for eating behaviour prediction is provided and the classification pipelines are explained. In Sect. 6.3.3, an overview of the experiment settings used in this work is provided. Results of the performed experiments are presented in Sect. 6.4. The obtained results are discussed in Sect. 6.5. The work is concluded in Sect. 6.6.

6.2   Dataset

The dataset used in this work was collected in the framework of the "ThinkSlim" study [7]. The dataset represents samples collected from 135 overweight participants over 8 weeks. The participants were asked to fill their data via the mobile app: (1) prior food intake and (2) at random, given the specified time intervals (as explained in [12]). In this work no distinction is made between the data collection and intervention stages of the Randomized Control Trial (RCT) used in the "ThinkSlim" study [7]. Additional characteristics of the dataset are described in [12].

A set of features extracted from the original dataset is described in Table 6.1. These features were selected based on the participants compliance in providing values for them, meaning that the features that were not filled regularly were excluded. Hereinafter, nominal and ordinal data are referred together as categorical data, and interval data is referred to as continuous data. From the data originally contained in feature F4, only the hour information was used.

Table 6.1 Dataset features used in the study. The used lists of assessed emotions and foods to select from are shown in Tables 6.2 and 6.3, respectively

Feature IDs	Feature name	Purpose	Data type	Variable type
F4	`date_saved`	Get DateTime of the sample	Nominal	Integer
F6	`circumstances`	Identify activity	Nominal	Integer
F7	`company`	Identify company size	Nominal	Integer
F8	`company_bond`	Quantify strength of a bond with the company	Interval	Double
F9	`craving`	Quantify desire to eat	Interval	Double
F17–24	`emotion_{...}`	Quantify emotional state	Ordinal	Integer
F30	`location_name`	Identify location	Nominal	Integer
F39	`specific_crave_names`	Identify desired foods	Set of nominal	{Integer}
F41	`specific_eat_names`	Identify consumed foods	Set of nominal	{Integer}

Table 6.2 Convention about the emotion types used in the study. Emotion names are given in English and Dutch (original language of the dataset)

Emotion type	Emotion names	
Positive	EN	Calm, Cheerful
	NL	Kalm, Opgewerkt
Negative	EN	Anxious, Angry, Nervous, Sad, Bored, Tired
	NL	Angstig, Boos, Gespannen, Verdrietig, Verveeld, Vermoeid

A list of assessed participant's emotions is shown in Table 6.2 together with their type attribution (positive/negative). Feature values related to these emotions were originally collected on the Visual Analogue Scale (VAS) scale (from 0 to 10) and were continuous. However, given the sparsity of the data, the feature values were converted to categorical. This step was performed following the motivation provided in [12]. Specifically, values corresponding to positive emotions were converted to three categories (Low/Mid/High) whereas values of negative emotions were converted to two categories (Yes/No). However, in contrast with [12], here zero-valued emotions were included in the category "Low".

The original data containing 46040 samples was preprocessed. First, 7069 duplicate entries were removed. Then, the data was time-sorted using feature F4. After that, 3274 samples related to drinking before going to bed were removed. Lastly, 184 samples with missing value for feature F41 (identifying food items consumed) were removed. The total size of the dataset after preprocessing was 35513 samples, out of which 20812 samples do not correspond to food intake.

Table 6.3 Convention about the food types used in the study. Emotion names are given in English and Dutch (original language of the dataset)

Food type		Food names
Healthy	EN	Nuts, Soup, Fruit, Pasta, Salad, Yoghurt, Potatoes, Rice dish, Cornflakes, Sandwiches
	NL	Nootjes, Soep, Fruit, Pasta, Salade, Yoghurt, Aardappelen, Rijstgerecht, Cornflakes, Boterhammen
Unhealthy	EN	Ice cream, Candy, Chips, Fries, Pizza, Muffin, Pastry, Cookies, Candy bar, Hamburger
	NL	IJs, Snoep, Chips, Frietjes, Pizza, Muffin, Gebak, Koekjes, Snoepreep, Hamburger

The individual names of food and their attribution to healthy and unhealthy food types were as shown in Table 6.3. The choice of the food types used in the "ThinkSlim" study was motivated by their presence in a typical Dutch diet [12, 23]. For classification, a label of unhealthy eating was assigned to a sample for which feature F41 contained any of the unhealthy food types (e.g. {Pasta} → "Healthy", {Pasta, Fruit} → "Healthy" and {Pasta, Muffin} → "Unhealthy"). This rather strict labelling of eating type was aimed at increasing the number of samples corresponding to unhealthy eating. Nevertheless, the dataset remained highly imbalanced in favour of healthy eating samples (1968 unhealthy eating and 12733 healthy eating samples). Moreover, the motivation of the study is to understand the triggers that lead users to make a choice which is less healthy than others. In that context, unhealthy eating refers to highly palatable foods, whereas healthy eating refers to the healthier choice.

After the EMA data was collected, the data logging compliance of the participants was checked. Specifically, the participants who logged less than two samples per day were considered non-compliant. This way, 80 participants (out of 135) were retained and the rest were excluded. For completeness, results before and after participants exclusion are presented in this work.

6.3 Analysis Methods

In this section, classification settings used in the experiments are defined and explained. Then, the tools used for data analysis and classification are introduced. Next, the outline of the performed experiments is presented. Finally, the methods used for interpretation and analysis of the trained models are detailed.

6.3.1 Classification Settings

Given that the dataset used in this work was collected with a goal of identifying unhealthy eating behaviour in adults, in this study, focus was on three prediction *targets* shown in Table 6.4. Namely, HU represents the primary target of interest. However, HUN and EAT prediction targets were used to test the classification performance for relevant prediction scenarios. HUN target is not only more general than HU, as no eating can also be predicted, but it also allows testing classification performance with a dataset more balanced w.r.t number of samples per class (for specific figures see Sect. 6.2). EAT target is of no direct value for the original goal of the study in which the dataset was collected, but it is deemed to be of indirect value, as it allows identification of food intake in general.

Two prediction *scenarios* were considered as shown in Table 6.5. Prediction at time t represents a setting in which analytics about person's behaviour is required. For instance, if it is necessary to verify for a given sample whether the participant's response was "truthful". On the other hand, prediction at time $t + 1$ represents a forecasting setting, in which the next behaviour sample is of interest. For instance, if it is desired to know whether a person is going to eat something unhealthy soon.

Two prediction *modes* were considered, as shown in Table 6.6. INPP_N mode is of particular interest w.r.t the study during which the dataset used was collected,

Table 6.4 Prediction targets considered

Code	Classes	Description
EAT	2 (eating, no eating)	Using data from all the features, predict whether the value of feature F41 indicates any food intake or not
HUN	3 (healthy, unhealthy eating, no eating)	Using data from all the features, predict whether the value of feature F41 contains any un/healthy food or if it contains no food at all
HU	2 (healthy, unhealthy eating)	Same as HUN, but all samples not corresponding to food intake are excluded from the dataset

Table 6.5 Prediction scenarios considered

(a) Predict target value at time t		(b) Predict target value at time $t + 1$	
Features at time 0	Target/label at time 0	Features at time 0	Target/label at time 1
...	...	...	...
Features at time t	Target/label at time t	Features at time t	Target/label at time $t + 1$
...	...	...	...
Features at time T	Target/label at time T	Features at time $T - 1$	Target/label at time T

Table 6.6 Prediction modes used

Code	Name	Description
IN	Incremental	Data is assumed to arrive sample by sample. No distinction between participants is made
INPP_N	Incremental per participant using history of length N	Data is assumed to arrive sample by sample. Classification models are trained individually per participant. To predict target/label at time t, previous N samples (taken at times $\{t - N, \ldots, t - 1\}, N \geq 1$) are used to form a feature vector. In addition, class labels from each of the history samples are included to the feature vector

as it allows predicting a target value with a constant number of preceding samples (history length) using the data of individual participant. IN mode represents a situation in which one common behaviour model is built using all the data available from the participants. The latter though does not comply with the perspective of personalization through individual behaviour modelling, and thus, serves as a reference.

6.3.2 Tools

Two tools were used for the off-line processing and analysis of the EMA dataset, namely: (1) KNIME Analytics Platform and (2) an own pipeline with custom classification routines written in C++. In KNIME Analytics Platform, the algorithms listed in Table 6.7 were applied to the dataset. These algorithms represent three different classes of classifiers: tree-based (DT, GBT, and RF), probabilistic/Bayesian (NB and BN), and neural network-based (MLP). Although this is by no means a complete coverage of possible classification methods (as for instance, maximum margin classifiers (Support Vector Machine (SVM)) and deep Artificial Neural Networks (ANNs) are not included), it should provide a realistic impression on the attainable classification performance using EMA data in the tested setting.

After the initial results were obtained with KNIME Analytics Platform, it was decided to implement an own set of classification routines based on the Naïve Bayes classifier. This choice was motivated by the simplicity of the method allowing direct interpretation of the results, its computational efficiency and a natural ability to accommodate for large sets of input training data in an incremental classification setting. More details about the own classification pipeline are provided in Sect. 6.3.2.1.

In the process of incremental training/testing, using both KNIME Analytics Platform and the own pipeline, classification results were not obtained for the sample collections containing only samples belonging to one class. Instead, a classifier was trained until the data contained samples belonging to more than one

Table 6.7 Algorithms used in KNIME Analytics Platform

Code	Description
BN	Bayesian network (entropy-based, max parents 2)
GBT	Gradient-boosted trees (max tree depth 4, 100 models, learning rate 0.1)
NB	Naïve Bayes
MLP	Multi-layer perceptron (5 hidden layers, 10 neurons per hidden layer, max 100 training iterations)
RF	Random forest (information gain-based, 100 models)
DT	Decision tree (Gini index-based, Minimum Description Length (MDL) pruning)

class, and only after that the classifier was tested. This was done to avoid introducing a bias in the classifier performance assessment, which would otherwise stem from correctly classifying the only existing class until the data contained more than one class.

All the classification accuracies reported in this work represent Balanced Accuracies (BAs), computed over K classes as

$$BA = \frac{\sum_{k=0}^{K} TPR_k}{K}, \tag{6.1}$$

where TPR_k is a True Positive Rate of the k-th class ($k \in K$).

6.3.2.1 Own Pipeline: Model Training and Testing

In case of conditionally independent features, the own pipeline implementation follows the same principle as the NB implemented in KNIME Analytics Platform. However, own pipeline permits the use of conditionally dependent features as described below. In the scope of this paper, the own pipeline was applied assuming the feature independence. Therefore, the results obtained using the NB implementation in KNIME Analytics Platform and the own pipeline are different mainly due to internal handling of data types.

Training

For all continuous variables, mean and variance are estimated per class. The estimation is performed incrementally. Such, the sample mean is computed as

$$\bar{x}_{t+1} = \frac{\sum_{t=1}^{t+1} x_t}{t+1}. \tag{6.2}$$

The sample variance is computed as

$$\sigma_{t+1} = \frac{(\sigma_t + \bar{x}_t^2) \cdot t + x_{t+1}^2}{t+1} - \bar{x}_{t+1}^2. \tag{6.3}$$

For all categorical variables, counts of feature value observations per class are stored

$$count(k, m) = \sum_{\substack{t=0 \\ k \in K \\ m \in M}}^{T} [c(t) = c_k] [f(t) = f_m], \tag{6.4}$$

where K classes and M categorical feature values are considered. (Each categorical feature has its own set of categorical values v_n, defined for $n \in N$. This means that for each feature, an own N exists. However, for compactness of notation, additional indexing is further avoided.) Here $[\cdot]$ is the Iverson conditional notation [24].

In addition, counts of observations of pairs of categorical feature values are stored. These counts can be formally expressed as

$$\text{count}(k, m, \hat{m}) = \sum_{\substack{t=0 \\ k \in K \\ m, \hat{m} \in M \\ m \neq \hat{m}}}^{T} [c(t) = c_k][f(t) = f_m][f(t) = f_{\hat{m}}], \tag{6.5}$$

where $f_{\hat{m}}$ is a categorical feature value other than f_m co-registered at time t. For instance, $f_m = $ "Alone" defines a company and $f_{\hat{m}} = $ "Home" defines a location.

Prediction

Prediction using the trained model is done as illustrated in Fig. 6.1. At time t a new sample of data $s(t)$ with M features (f_m for $m \in M$) is recorded. Given a set of K classes, for each class c_k a probability is computed following the Bayes rule as

$$P(c_k|s(t)) = \frac{P(s(t)|c_k)P(c_k)}{P(s(t))} \propto P(c_k)P(s(t)|c_k) \propto P(c_k) \prod P(f_m|c_k), \tag{6.6}$$

where $P(c_k|s(t))$ is a posterior conditional probability of class c_k given sample $s(t)$, $P(s(t)|c_k)$ is a conditional probability of sample $s(t)$ given class c_k, $P(c_k)$ is a prior probability of class c_k, $P(s(t))$ is a total probability of sample $s(t)$, and $P(f_m|c_k)$ is a conditional probability of the m-th feature value given class c_k.

The interrelations between the feature values can be captured by computing conditional probabilities of feature value pairs. Then, the posterior probability of class c_k is be computed as

$$P(c_k|s(t)) \propto P(c_k) \prod_{m \in M} P(f_m|c_k) \prod_{\substack{m, \hat{m} \in M \\ m \neq \hat{m}}} P(f_m, f_{\hat{m}}|c_k), \tag{6.7}$$

where $P(f_m, f_{\hat{m}}|c_k)$ is a conditional probability of a pair of values of the m-th and $\hat{m}$-th features given class c_k.

Once the posterior probabilities are computed for all K classes, the class with the highest posterior probability is selected as the predicted label for sample $s(t)$. The same procedure is repeated for all unseen samples.

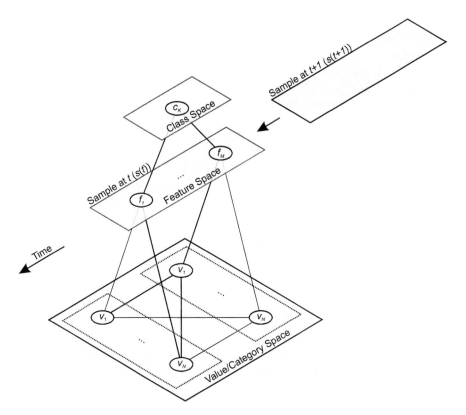

Fig. 6.1 Prediction model used in own pipeline. For a sample at time t ($s(t)$), probability of class c_k ($k \in K$) is computed as a product of conditional probabilities of M feature values of the sample (e.g. $P(c_k|s(t)) \propto P(f_1 = v_1|c_k) \cdot \ldots \cdot P(f_M = v_N|c_k)$). In addition, conditional probabilities of pairs of feature values can be used to take into account interrelations between the feature values (e.g. $P(f_1 = v_1, f_M = v_N|c_k)$). Thicker lines denote higher probability values. Values are grouped by a dotted rectangle in the value space based on their attribution to a particular feature

6.3.3 Experiments

The legend of experiments performed is shown in Table 6.8. The experiments with KNIME Analytics Platform were performed only in the INPP_1 prediction mode. This was due to the platform limitations w.r.t incremental learning. Namely, at the moment, models cannot be incrementally trained in KNIME Analytics Platform. This means, that predicting a label for each new sample requires training of a new model from scratch using all the currently available data. The fact that the models cannot be reused makes running the incremental experiments (especially in IN mode) so computationally expensive, that the experiments do not produce results within a reasonable time. (For instance, an attempt to obtain results with KNIME Analytics Platform using MLP in IN mode took 100 h but was still not finished.)

Table 6.8 Experiments' legend. For explanations of codes of prediction targets and modes see Tables 6.4 and 6.6, respectively

Experiment ID	Prediction target	Predict at time	Dataset size (samples) [a]	Prediction mode	Tool
K.X	EAT	t	34891		
K.XI	HU	t	13611		
K.XII	HUN	t	34944	INPP_1	KNIME Analytics Platform
K.XIII	EAT	$t+1$	34751		
K.XIV	HU	$t+1$	13579		
K.XV	HUN	$t+1$	34806		
O.I	EAT	t	35206		
O.II	HU	t	13738		
O.III	HUN	t	35260	See Table 6.11	Own pipeline
O.IV	EAT	$t+1$	35066		
O.V	HU	$t+1$	13705		
O.VI	HUN	$t+1$	35122		

[a] A number of classification decisions made when no participants were excluded (135 participants included). For experiments with own pipeline the sizes from INPP_1 mode are reported

The results obtained with KNIME Analytics Platform provide a reference for the results obtained with the own pipeline.

6.3.4 Model Interpretation and Analysis

Along the incremental training process, prediction models were saved once a week for each of the participants. This frequency of saving models was selected as it was deemed to be minimally sufficient to observe changes in the participant's behaviour. The prediction models trained using the own pipeline were processed in two steps:

1. Each of the models is interpreted as follows:

   - Every model contains information used to classify among a set of classes. A number of classes K is identified, and for each of the classes c_k ($k \in K$) a set of categorical features f_m ($m \in M$) used for the prediction of the class is extracted.
   - Data related to categorical and continuous features is analysed independently (the details are provided is separate sections below).

2. Using the obtained model interpretations, the selected models are compared with each other (as described in Sect. 6.3.4.4).

6.3.4.1   Analysis of Categorical Features

Assuming that a class c_k is considered (e.g. class "eating healthy"), for each categorical feature value $f_{\hat{m}}$, $\hat{m} \in M$ (e.g. for category "Home" of the `location_name` feature), normalized category frequencies are obtained as

$$\text{freq}^{c_k}_{f_{\hat{m}}} = \frac{\text{count}(k, \hat{m})}{\sum_{k \in K} \text{count}(k, m)}, \tag{6.8}$$

where count$(\cdot)$ values are computed using Eq. (6.4). The obtained normalized category frequencies are then sorted for each of the predicted classes.

To obtain values characterizing an ability of the model to discriminate among the predicted classes, all possible pairs of classes to discriminate from are created. For every pair of classes (e.g. c_k versus $c_{\hat{k}}$), for all shared categorical features and for all corresponding categories, the discrimination values are found as

$$\text{freq}^{\text{disc}}_{f_{\hat{m}}} = \text{freq}^{c_k}_{f_{\hat{m}}} - \text{freq}^{c_{\hat{k}}}_{f_{\hat{m}}}. \tag{6.9}$$

Note that $\text{freq}^{\text{disc}}_{f_{\hat{m}}}$ is in range $[-1, 1]$.

6.3.4.2   Analysis of Continuous Features

For a continuous feature (e.g. `craving`), distribution of its values between the classes characterizes how well the feature can be used to discriminate between the classes. Specifically, the role of the $\hat{m}$-th feature in discriminating between two classes c_k and $c_{\hat{k}}$ is determined as Fisher's discrimination criterion [25]:

$$J^{\text{disc}}_{\hat{m}} = \frac{s^{\text{between}}_{\hat{m}}}{s^{\text{within}}_{\hat{m}}} = \frac{w^{c_k} w^{c_{\hat{k}}} d^2(\bar{x}^{c_k}_{\hat{m}}, \bar{x}^{c_{\hat{k}}}_{\hat{m}})}{\sigma^{c_k}_{\hat{m}} + \sigma^{c_{\hat{k}}}_{\hat{m}}} = \frac{\sum_{k \in K} w^{c_k} d^2(\bar{x}^{all}_{\hat{m}}, \bar{x}^{c_k}_{\hat{m}})}{\sum_{k \in K} \sigma^{c_k}_{\hat{m}}}, \tag{6.10}$$

where $s^{\text{between}}_{\hat{m}}$ and $s^{\text{within}}_{\hat{m}}$ are the between- and within-class variances w.r.t values of the $\hat{m}$ feature, respectively, w^{c_k} is a weight of class c_k in the dataset (i.e. frequency of samples belonging to class c_k), $\bar{x}^{c_k}_{\hat{m}}$ and $\sigma^{c_k}_{\hat{m}}$ are a mean and a variance of the $\hat{m}$-th feature values from the samples belonging to class c_k respectively, $d^2(\cdot)$ is a Squared Euclidean Distance (SED) $(d^2(\mathbf{x}, \mathbf{y}) = \sum_{i=0}^{N}(x_i - y_i)^2)$ and $\bar{x}^{all}_{\hat{m}}$ is an overall mean of the $\hat{m}$-th feature values irrespective of the sample's class attribution. This means that the feature which maximizes a scatter of values between the classes and minimizes a scatter of values within the classes has a high discriminatory capacity for the given classes.

The overall mean $\bar{x}^{all}_{\hat{m}}$ (used above) and the overall variance $\sigma^{all}_{\hat{m}}$ (will be used below) can be efficiently estimated from the class-related means $\bar{x}^{c_k}_{\hat{m}}$ by pooling:

$$\bar{x}_{\hat{m}}^{all} = \frac{\sum_{k \in K} w^{c_k} \bar{x}_{\hat{m}}^{c_k}}{\sum_{k \in K} w^{c_k}}$$

$$\sigma_{\hat{m}}^{all} = \frac{\sum_{k \in K} (w^{c_k} - 1)\sigma_{\hat{m}}^{c_k}}{\sum_{k \in K} (w^{c_k} - 1)}. \qquad (6.11)$$

Attempting to have the $J_{\hat{m}}^{disc}$ values normalized, instead of using the SED, the normalized SED can be used:

$$d^2(\bar{x}_{\hat{m}}, \bar{x}_{\hat{m}}^{c_k}) = \frac{0.5\sigma_{\hat{m}}^{all-c_k}}{\sigma_{\hat{m}}^{all} + \sigma_{\hat{m}}^{c_k}}, \qquad (6.12)$$

where $\sigma_{\hat{m}}^{all-c_k} = \sigma_{\hat{m}}^{all} + \sigma_{\hat{m}}^{c_k} - 2\text{cov}(all, c_k)$. However, a direct computation of $\text{cov}(all, c_k)$ is computationally complex, thus, the Cauchy–Schwarz inequality ($|\text{cov}(all, c_k)| \leq \sqrt{\sigma_{\hat{m}}^{all}\sigma_{\hat{m}}^{c_k}}$) can be used to estimate the covariance. Nevertheless, although the use of the normalized SED allows reducing the scatter of $J_{\hat{m}}^{disc}$, the normality of $J_{\hat{m}}^{disc}$ is not guaranteed as in a general case the class-related variances $\sigma_{\hat{m}}^{c_k}$ are not normalized. Specifically, variance of the feature values for a class can be small (for instance, if the number of samples belonging to the class is small) or zero (for instance, if there is only one sample belonging to the class or if all the samples are the same). Therefore, it is suggested to impose a cap of one on the $J_{\hat{m}}^{disc}$ values estimated using the normalized SED, such that $J_{\hat{m}}^{disc}$ is in [0, 1] range.

6.3.4.3 Interpretation of the Resulting Values

The class discrimination values obtained for continuous and categorical features cannot be mixed and are analysed separately. This means that, for instance, feature company cannot be analysed w.r.t the company_bond feature.

The closer the value of $\text{freq}_{f_{\hat{m}}}^{c_k}$ is to 1, the higher is the probability of observing class c_k if $f_{\hat{m}}$ is recorded. This means, that this value can be perceived as a "weight" of the category in making a prediction decision in favour of the given class.

The closer the value of $\text{freq}_{f_{\hat{m}}}^{disc}$ is to -1, the higher is the probability of observing class $c_{\hat{k}}$ when the category $f_{\hat{m}}$ is recorded. On the other hand, the closer the value of $\text{freq}_{f_{\hat{m}}}^{disc}$ is to 1, the higher is the probability of observing class c_k when the category $f_{\hat{m}}$ is recorded. This means, that those values identify which categories of the categorical features contribute the most to deciding about observing one of the two classes. If $\text{freq}_{f_{\hat{m}}}^{disc}$ is close to zero, it means that category $f_{\hat{m}}$ has a "similar" impact on a probability of observing either of the two classes, and thus is not "important" for the prediction.

The closer the value of $J_{\hat{m}}^{disc}$ is to zero, the lower is the impact of the feature in discriminating between the classes. Therefore, the features with the values of $J_{\hat{m}}^{disc}$ closer to one should be retained.

6.3.4.4   Model Comparison

At any time t, for every participant a limited amount of data is available. Therefore, it can be said that for every participant there exists a prediction model which is valid at time t. These prediction models are learnt per participant individually. Once an additional piece of data becomes available for any of the participants, the time is incremented for all the models. This means, that if, for instance, at time $t - 1$ new data is available for participant $n-1$, then a new model is trained for participant $n-1$ at time $t-1$ and the models of all other participants remain valid at time $t-1$ without changes.

The process of model training and comparison is shown in Fig. 6.2. For compactness, data and models are shown only for two users. The models trained as explained in Sect. 6.3.2.1 can be compared using the following comparison types

I. The models of two different users corresponding to the same time can be compared. This way is possible to identify relatedness between the users based on their behaviour patterns at a set time. This can be useful for user grouping. For instance, when a peer should be assigned to a user as a companion in a CBT.
II. The models of a user corresponding to different times can be compared. For example, comparing models obtained after 2 and 7 weeks of an EMA study. This way, a behaviour change is identified, and the overall degree of the change is quantified using the aforementioned distance metrics.
III. The models of two different users obtained at two different times can be compared. This type of comparison allows identification of similar behaviour patterns across participants at different times. Overall, this is a generalization of the comparison types I. and II.
IV. The model of a user at a given time can be compared to an aggregate model of other users at that time. The aggregated model can be obtained, for instance, as an average of models of other users. The aggregated model can be perceived to capture behaviour of an average user. This can be particularly useful when there is a need to establish a data-driven baseline of behaviour, for instance, to provide feedback to the users relative to the group baseline. For example, if a participant significantly deviates from the group behaviour in a number of snacking moments per day, then a targeted feedback can be issued.

Only the models trained to predict for the same set of classes can be compared. Moreover, in a comparison, it must be ensured that the values corresponding to the same features are compared (for instance, company_bond feature of one model is compared with the same feature of another model).

Specifically, the model comparison is performed using the values described in Sect. 6.3.4.3. The $\text{freq}^{c_k}_{f_{\hat{m}}}$ (and $J^{disc}_{\hat{m}}$) values form a vector uniquely identifying a prediction model. Hence, a distance between the vectors can be computed to see how similar are the prediction models and the user behaviours which the models were trained on.

Euclidean and cosine distance metrics are computed to identify the degree of model/behaviour similarity. Cosine distance is of particular interest for analysis of

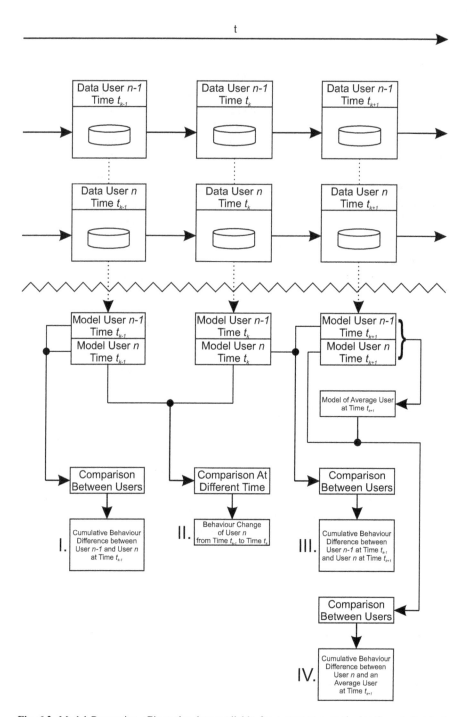

Fig. 6.2 Model Processing. Given the data available for a user at a particular time, individual models are obtained per user. Four types of model comparison are suggested as described in Sect. 6.3.4.4

EMA data, as it allows identifying a difference in "directions" between the participant behaviours. Comparing the exact parameters of the models using Euclidean distance is in turn useful when an exact match in the behaviour proportions captured in the models is desired. The latter is expected to be particularly useful in the type IV comparison with a reference model, setting a goal for the participant's behaviour.

For instance, comparing the models obtained during an EMA study with the model obtained at the end of the study allows observing the convergence in model training (more details are provided in Sect. 6.4.3). The convergence may identify a suitable time to start an intervention (an EMI). A threshold on the distance(s) can be set to acknowledge the convergence.

6.4 Results

The results of this work are presented in the following three blocks. First, the results obtained with KNIME Analytics Platform are summarized in Sect. 6.4.1. Then, the results obtained with the own classification pipeline are presented in Sect. 6.4.2. Finally, several examples of model analysis and comparison are provided in Sect. 6.4.3.

6.4.1 Experiments with KNIME Analytics Platform

The results obtained from the experiments with KNIME Analytics Platform without exclusion of participants are summarized in Table 6.9. For all the algorithms, in all the experiments, the mean BA values were above chance. It can be seen that predicting a target at time $t + 1$ was more challenging than at time t. Predicting EAT target led to the highest BA values, and predicting HUN target led to the lowest BA values. None of the algorithms consistently outperformed others in all the experiments. However, when predicting a target at time $t + 1$, GBT performed better than other algorithms. Correct prediction of the dominant class of no eating led to the prediction of HUN target having the larger margin between the demonstrated BA values and the corresponding chance level (33.3 %) than the prediction of HU target (chance level of 50 %) both when predicting at time t and at $t + 1$.

When the participant exclusion was done, as described in Sect. 6.2, the results obtained with KNIME Analytics Platform were refined and are shown in Table 6.10. It can be seen that the participant exclusion led to marginal improvements in the BA values. More importantly, since the results of non-conforming participants were excluded, a reduction in the SD values was observed. The reduction in SD values was the largest in experiments K.XIII and K.XV. However, this reduction was not even across the algorithms and experiments, and in some cases the opposite was observed (for instance, see GBT in K.XII or NB in K.X). Similar to the

Table 6.9 Mean BA (%) values across algorithms for experiments with KNIME Analytics Platform, without exclusion of participants (135 participants included). For the description of the experiment conditions see Table 6.8. For the description of the algorithms see Table 6.7. Highest BA values and lowest SD values per experiment are marked in bold

Experiment ID			Algorithm code					
			BN	GBT	NB	MLP	RF	DT
K.X	Micro-average	Mean	73.8981	74.9293	76.2610	72.9295	76.1787	**77.9832**
	Macro-average	Mean	68.9964	69.6884	70.2533	67.3135	70.0070	**72.9499**
		SD	**11.4166**	12.4249	13.1965	11.8715	12.9949	11.8677
K.XI	Micro-average	Mean	62.3419	**69.2225**	67.4844	64.3286	62.4810	68.6808
	Macro-average	Mean	59.9984	67.1906	65.6207	61.3503	60.0851	**69.0474**
		SD	14.9543	16.3338	**13.9535**	16.6390	14.6218	17.4741
K.XII	Micro-average	Mean	53.5665	57.9644	**58.8995**	54.0631	56.5150	57.3104
	Macro-average	Mean	48.9024	52.7931	**53.5583**	49.3488	51.3953	53.3222
		SD	11.8171	13.2191	12.5308	**11.6290**	12.4647	13.3831
K.XIII	Micro-average	Mean	59.4071	61.2225	59.6438	61.0591	**61.7822**	61.7819
	Macro-average	Mean	55.1730	**57.5146**	56.1947	56.4148	56.5961	56.2361
		SD	7.1543	**6.6721**	7.1261	7.3463	7.1014	7.3575
K.XIV	Micro-average	Mean	54.3346	**55.9656**	55.8611	54.3366	54.0882	54.6483
	Macro-average	Mean	53.6562	**55.9327**	54.1624	54.2543	54.2845	54.7748
		SD	12.2840	13.5056	12.1622	13.0555	**12.1015**	12.7492
K.XV	Micro-average	Mean	40.7730	**42.8014**	41.6927	41.5904	42.5221	41.5091
	Macro-average	Mean	38.1588	**40.4863**	38.8830	39.0557	39.5094	38.7787
		SD	7.6432	8.4349	**6.3723**	7.8048	8.0058	7.7549

results obtained without participant exclusion (shown in Table 6.9), no algorithm consistently outperformed others in the conducted experiments.

6.4.2 Experiments with Own Pipeline

The results obtained from the experiments with own pipeline without exclusion of participants are summarized in Table 6.11. The results obtained in the INPP_1 mode by design most closely resemble the ones obtained for NB algorithm used in KNIME Analytics Platform (see Table 6.9 for reference). It can be seen that predicting a target at time t in the IN mode resulted in the consistently higher BA values than in the other prediction modes. Moreover, the use of extended history length, in the INPP_2 and INPP_5 modes, led to a reduction in the BA and SD values. However, in experiments O.IV and O.VI in INPP_2 mode the opposite was observed.

The development of the BA values over samples is visualized per participant for experiments O.II and O.V in Figs. 6.3 and 6.4, respectively. It can be seen that in experiment O.II the BA values obtained do not plateau at the end of data collection,

Table 6.10 Mean BA (%) values across algorithms for experiments with KNIME Analytics Platform, with exclusion of participants (80 participants included). For the description of the experiment conditions see Table 6.8. For the description of the algorithms see Table 6.7. Highest BA values and lowest SD values per experiment are marked in bold

Experiment ID			Algorithm code					
			BN	GBT	NB	MLP	RF	DT
K.X	Micro-average	Mean	74.9609	75.8483	76.7619	74.2079	77.0553	**79.4886**
	Macro-average	Mean	73.0988	74.0024	69.7437	71.5808	74.6756	**77.3878**
		SD	8.5059	**7.8530**	14.8930	8.2544	8.5229	8.2904
K.XI	Micro-average	Mean	63.1462	**69.5709**	68.8903	64.7700	63.2218	69.3985
	Macro-average	Mean	59.8829	67.1376	67.3980	62.7543	60.4094	**67.9828**
		SD	**11.9158**	13.2723	13.7045	12.9061	12.2413	15.8544
K.XII	Micro-average	Mean	54.7315	58.6290	**60.4180**	54.8650	57.4249	58.7011
	Macro-average	Mean	52.1798	51.9673	**57.3788**	51.9892	54.3639	56.8104
		SD	**8.5507**	14.8886	9.6720	8.6573	9.6733	10.4301
K.XIII	Micro-average	Mean	59.4376	61.5114	59.5801	61.5428	61.8314	**62.0851**
	Macro-average	Mean	56.3091	**58.6620**	57.0321	57.8265	57.9359	58.4518
		SD	**4.4620**	4.6283	4.7761	5.7458	4.9374	5.8948
K.XIV	Micro-average	Mean	54.5834	56.3849	**56.5943**	54.4837	54.5397	55.2516
	Macro-average	Mean	53.3285	**55.4088**	54.4301	53.0690	53.1556	54.6827
		SD	9.8673	11.2485	10.8339	10.2546	**9.1068**	11.2673
K.XV	Micro-average	Mean	40.7756	**43.0951**	40.8581	42.0030	41.8854	41.7457
	Macro-average	Mean	38.3376	**40.4473**	38.7129	39.4824	39.7663	39.1795
		SD	**4.2560**	4.7825	7.4424	4.8234	9.8752	4.8360

and that the personal behaviour model training would continue if the data was collected further. On the other hand, in experiment O.V, for most of the participants, the plateau is reached at ≈150 samples. It can be seen that in both experiments the phase of active model training lasts at least for the first 50 samples. For this reason, the results are visualized only for 64 participants for whom more than 50 samples were classified. The two-fold higher SD in experiment O.II compared to the one obtained in experiment O.V is also visible when comparing the two plots.

6.4.3 Model Interpretation and Analysis

A comparison between the models trained for participant pp065 over a period of 8 weeks is shown in Fig. 6.5 as an example of a type II comparison described in Sect. 6.3.4.4. Participant pp065 was picked randomly among the participants who logged the most of data after participant exclusion. It can be seen that when comparing the model obtained in week eight with the models obtained in the previous weeks, see Fig. 6.5, the distance between the compared models gradually decreased for both prediction targets (healthy and unhealthy eating). This means

Table 6.11 Mean BA (%) values across prediction modes for experiments with own pipeline, without exclusion of participants (135 participants included). For the description of the experiment conditions see Table 6.8. For the description of the prediction modes see Table 6.6. Highest BA values and lowest SD values per experiment are marked in bold

Experiment ID			Prediction mode code			
			IN	INPP_1	INPP_2	INPP_5
O.I	Micro-average	Mean		75.8968	74.5371	69.5364
	Macro-average	Mean	76.7012[a]	69.8268	68.3896	62.6129
		SD		13.1627	12.2634	11.8576
O.II	Micro-average	Mean		66.4217	63.2799	57.9837
	Macro-average	Mean	71.6037[a]	61.4077	57.8896	53.5591
		SD		10.4503	11.1327	8.7541
O.III	Micro-average	Mean		58.8039	55.5748	49.3046
	Macro-average	Mean	60.8411[a]	52.7891	49.6556	43.6607
		SD		12.1900	11.0198	9.5810
O.IV	Micro-average	Mean		56.4608	**58.1442**	56.0929
	Macro-average	Mean	N/A[b]	53.3052	**53.8914**	51.5119
		SD		6.2321	7.1387	**6.2213**
O.V	Micro-average	Mean		**55.1493**	54.0989	53.4476
	Macro-average	Mean	N/A[b]	**51.6733**	51.3217	51.0044
		SD		7.1760	6.0173	**4.7079**
O.VI	Micro-average	Mean		41.4747	**41.7302**	39.5337
	Macro-average	Mean	N/A[b]	**38.4120**	38.1273	36.1687
		SD		**5.6170**	6.2567	5.6645

[a]Obtaining micro- and macro-averages in the IN mode is not possible
[b]Obtaining results in the IN mode does not make sense

convergence of the model parameters over time. The higher the slope of the distance curves, the faster the convergence. If over time the distance is no longer decreasing, then no model alterations w.r.t a given prediction target are observed. The latter can occur, for instance, when none of the samples belonging to a given target was registered in the monitored period of time.

If the distance between the models increases, this means that a different behaviour was exhibited by the participant in the monitored time period. This can be observed comparing the models trained for participant pp065 in weeks 3 and 4 for healthy eating class and in weeks six to eight for unhealthy eating class, as shown in Fig. 6.5. When such an event occurs, it might be of interest to analyse the participant's behaviour in this time period more closely to see whether the change in behaviour is desired. For instance, if the participant started to eat unhealthy snacks before going to bed, this might not be desired and an action might have to be taken to prevent this from developing into a habit.

An example of a type III model comparison, as described in Sect. 6.3.4.4, is shown in Fig. 6.6 for participants pp065 and pp161. Participant pp161 was picked randomly among the participants who logged the most of data after participant

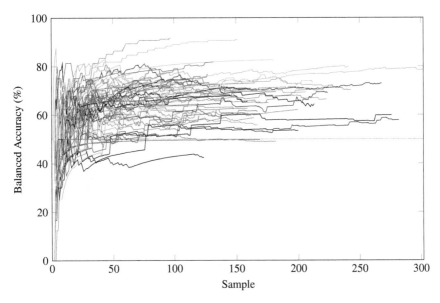

Fig. 6.3 Development of BA values per participant across samples for experiment O.II in INPP_1 prediction mode. Only samples for participants for whom more than 50 samples were classified (64 out of 80 participants after exclusion). Chance level is shown as a grey horizontal line

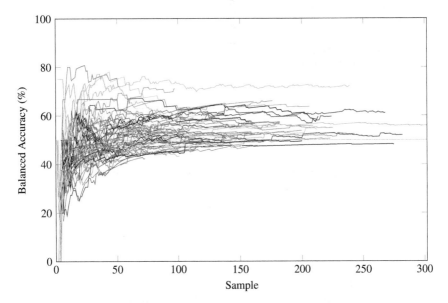

Fig. 6.4 Development of BA values per participant across samples for experiment O.V in INPP_1 prediction mode. Only samples for participants for whom more than 50 samples were classified (64 out of 80 participants after exclusion). Chance level is shown as a grey horizontal line

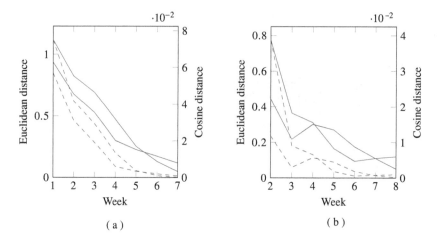

Fig. 6.5 Convergence of models trained over time for participant pp065 in experiment O.II. In **(a)**, the model at week eight is compared with the model at week N. In **(b)**, the model at week N is compared with the model at week $N - 1$. Distance between the models describing healthy eating is shown in blue and unhealthy eating is shown in red. Solid lines refer to Euclidean distance (left axis) and dashed lines refer to cosine distance (right axis)

Fig. 6.6 Distances between the models of participants pp065 and pp161 in experiment O.II (a type III comparison, as described in Sect. 6.3.4.4). Model describing healthy eating is shown in blue and unhealthy eating is shown in red. Solid lines refer to Euclidean distance (left axis) and dashed lines refer to cosine distance (right axis)

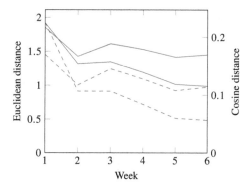

exclusion. Participant pp161 logged data for 6 weeks only, thus the models of the first 6 weeks for participant pp065 were used for comparison. It can be seen that after week two, the models became more similar in describing behaviour leading to healthy eating and less similar in describing behaviour leading to unhealthy eating. This trend persisted over the remaining weeks. Moreover, it can be seen that overall, these two participants had more in common in their behaviour leading to healthy eating than in the behaviour leading to unhealthy eating (no significant difference in cosine ($p = 0.3968$, $\alpha = 0.05$) and Euclidean distances ($p = 0.1372$, $\alpha = 0.05$)). As anticipated, the models of the two participants are significantly less similar than the models obtained from participant pp065 across the same 6-week period (compare cosine distances in Figs. 6.5 and 6.6 for healthy eating ($p = 0.0011$, $\alpha = 0.05$) and for unhealthy eating ($p = 1.6025 \times 10^{-7}$, $\alpha = 0.05$)).

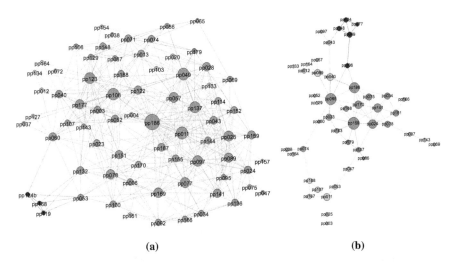

(a)   (b)

Fig. 6.7 Clusters of models obtained in experiment O.II based on cosine distance between the models predicting: (**a**) healthy and (**b**) unhealthy eating. Models compared here were trained using all the samples available for the participant. Only the models of 80 participants after exclusion were analysed. Node size reflects degree. Node colour represents the cluster assignment. Edges represent cosine distances between the models of a pair of participants. Only the distances below 0.1 included. Participants with degree of zero and disconnected groups of two participants only are not shown

Using the same approach for comparing the models between participants, the participants behaviour was clustered using Leiden algorithm [26] in Gephi [27]. The resulting clusters are shown in Fig. 6.7. It can be seen that the healthy eating class was characterized by one large strongly connected cluster, which means that the participants in that cluster shared their healthy eating behaviour patterns. Minority clusters for the healthy eating class contained 2–3 participants only. Notable is the high degree (a number of connected edges) of participant pp186 w.r.t the healthy eating class. This means, that the behaviour of this participant was similar to the behaviours of other participants, and that participant pp186 well "represented" the group's healthy eating behaviour. For the unhealthy eating class however, no single dominating cluster was found. The resulting clusters were less strongly connected internally and were more distant from each other than for the healthy eating class. This suggests that the unhealthy eating class was associated with distinct behaviour patterns exhibited by groups of participants. Though, the behaviours were not different enough to form clusters isolated from each other, which means that the behaviours had shared traits.

An example of a type IV comparison between the combined model ("model of an average user") obtained from 80 participants after exclusion at the end of the experiment and the models of participant pp065 obtained at different weeks in experiment O.II is shown in Fig. 6.8. It can be seen that the distances between the models w.r.t unhealthy eating followed a flat trend across

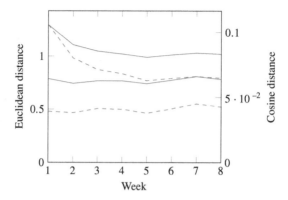

Fig. 6.8 Distances between the combined model obtained from 80 participants after exclusion at the end of the experiment and the models of participant pp065 obtained at different weeks in experiment O.II (a type IV comparison, as described in Sect. 6.3.4.4). Model describing healthy eating is shown in blue and unhealthy eating is shown in red. Solid lines refer to Euclidean distance (left axis) and dashed lines refer to cosine distance (right axis)

all the 8 weeks. On the other hand, w.r.t healthy eating, the distances between the models were decreasing until week five, after which a similar flat trend emerged. The distances between the models w.r.t healthy eating were consistently and significantly higher than the distances w.r.t unhealthy eating across all the 8 weeks (significant difference in cosine ($p = 2.0791 \times 10^{-5}$, $\alpha = 0.05$) and Euclidean distances ($p = 1.1267 \times 10^{-6}$, $\alpha = 0.05$)). The observed dynamics shows that the underlying behaviour of participant pp065 was consistently in line with an "average" behaviour of the other participants. Moreover, the healthy eating behaviour tended to converge to the behaviour of the other participants, unlike the unhealthy behaviour.

The contributions of individual categorical feature values in deciding for a predicted class, computed per class using Eq. (6.8), can be visualized per participant as shown in Fig. 6.9 for participant pp060 in experiment O.V in INPP_1 mode. Participant pp060 was picked randomly among the participants who logged the most of data not taking participant exclusion into account. In this graph, the nodes located closer to the corresponding class label (healthy or unhealthy eating) have a higher contribution to predicting this class than the other class. For the two-class case shown here, the dipole-like graph is obtained, and the nodes located on either side of the dipole (a virtual "border" is formed by a normal at the centre of a line connecting the nodes representing the two classes) have a higher contribution to the corresponding class.

An example of applying correlation analysis for discovery of relatedness between the categorical feature values given in class is shown in Fig. 6.10 for participant pp060 in experiment O.V in INPP_1 mode for unhealthy eating class. This perspective demonstrates combinations of factors which are prominent when the participant is going to eat unhealthy (at time $t + 1$). It can be seen that three groups of correlated categorical feature values are present: (1) the large group in the centre

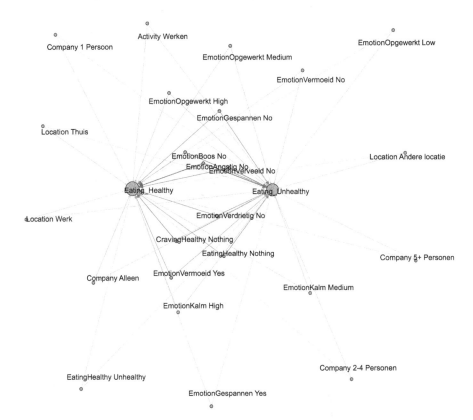

Fig. 6.9 Contributions of feature categories in deciding for a prediction target in experiment O.V in INPP_1 mode for participant pp060. Node size represents degree. Edge thickness represents contribution weight (the thicker, the higher the contribution). Nodes with contribution below 0.1 are not shown for compactness

of the graph, which includes highly related emotional components; (2) a group at the top of the graph, which includes negative emotions; and (3) a group of two components on the right side of the graph, which specifically describes the situation of being bored using an e-device.

6.5   Discussion

The results obtained with the used EMA dataset demonstrated the different levels of complexity in the selected prediction targets. Firstly, predicting a class value at time $t + 1$ was significantly more challenging than predicting the same value at time t in all experiments both for KNIME Analytics Platform and the own pipeline (specific significance test results are omitted for compactness). This means that the consistency in samples describing the current activity was higher compared

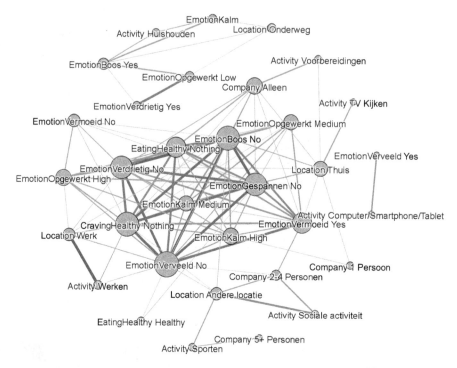

Fig. 6.10 Correlations between the feature categories when predicting for unhealthy eating class in experiment O.V in INPP_1 mode for participant pp060. Node size represents degree. Edge thickness represents the degree of correlation (the thicker, the higher the correlation). Only edges with $r \geq 0.5$ are shown for compactness

to the consistency in samples describing the future activity. Therefore, determining a trend that a person, for instance, tends to eat unhealthy in a company at work is easier than determining a trend that, for instance, if the person eats unhealthy in a company at work than the person will eat unhealthy after that.

However, the reasons for this observation are not clear. This might be due to the increased user state space which is being modelled when predicting at time $t+1$ and the consequential increase in the number of behaviour samples required to capture the transitions in the state space. Alternatively, this can be caused by the increased number of inconsistencies in samples which hinder prediction performance. Solving both of the mentioned potential causes is possible by obtaining more samples of the user behaviour; however, the costs and time necessary for the extensive data collection should be considered.

Although the prediction of HUN target led to significantly lower BA values than the prediction of HU target (e.g. when comparing micro-average BA values obtained in experiments K.XI and K.XII ($p = 1.1713 \times 10^{-4}$, $\alpha = 0.05$)), for the HUN target, the BA values were more significantly different from the 3-class chance level of 33.3 % (e.g. for micro-average BA values obtained in experiment K.V 95 % Confidence Interval (CI) [41.2248, 42.4047]) than the BA values obtained when predict-ing for the HU target with the chance level of 50 % (for micro-average BA values obtained in experiment K.IV 95 % CI [54.2112, 55.5336]). Moreover, prediction for the HUN target resulted in significantly lower variance of the obtained BA values than the prediction done for the HU target (e.g. when comparing SD values obtained in experiments K.XI and K.XII ($p = 4.9963 \times 10^{-4}$, $\alpha = 0.05$)). Therefore, the HUN prediction target can be useful in a situation when errors in predicting no eating versus healthy/unhealthy eating are tolerable (e.g. when given an unhealthy eating sample, erroneously predicting no eating is less critical than predicting healthy eating).

Exclusion of participants based on their compliance to the data entry led to marginal improvements in the obtained BA values and to a reduction of variance across the participants. Nevertheless, the effect of participant exclusion was not significant (specific significance test results are omitted for compactness).

The use of the INPP_2 and INPP_5 prediction modes in own pipeline did not lead to a consistent performance improvement compared to the INPP_1 mode. It merely resulted in the reduction of variance in the BA values across the participants. The decrease in the mean BA values can be explained by the increased dimensionality of the feature space without an increase in the number of training samples. Given this observation, performing an additional feature selection step might help limiting the number of used features.

The results obtained with KNIME Analytics Platform indicated that GBT algo-rithm bears the highest potential for predicting a target at time $t + 1$ in the incremental setting among the tested algorithms. This observation complements the suggestions made previously w.r.t the use of boosted trees in a CV setting [14]. It is unclear however whether the use of a set of weighted trees can provide an adequate level of explanation of the model functioning, as the set should be examined as a complete entity at once and the notion of weights can be misleading. Nevertheless, GBT algorithm is to be considered in the future research.

6.5.1 Is Personalization Always Necessary?

The results obtained with the own pipeline, shown in Table 6.11, indicate that the mean BA values obtained in the INPP_{...} prediction modes with the tested group of participants are lower than the results obtained in the IN prediction mode. This means that training personalized prediction models individually for each participant did not lead to an improved classification performance on average. Therefore, the use of a single "global" model of user behaviour would be more accurate on average.

However, as shown in Fig. 6.4, for some participants the use of the personalized prediction models built based on their own behaviour samples demonstrated higher BA values, towards the end of the data collection process, than the group mean and than the ones obtained in the IN prediction mode (without a distinction across participants). This means that the use of the personalized models for these participants would be more beneficial than the use of a "global" model.

Given these two observations, it is suggested to incrementally train both a "global" and the personalized prediction models. Testing the "global" model would allow setting a classification performance baseline. Then, the performance of a personalized prediction model of a particular user can be compared with this baseline. If the personalized model is shown to outperform the "global" model, then the personalized model is used, otherwise, the "global" model is used.

This approach does not put an additional computational burden as only a mere comparison between the model BA values is required. However, in case if, for instance, a Multi-Layer Perceptron (MLP) classifier is used, an incremental training of a "global" model given a potentially large dataset might lead to a considerable increase of the training time. On the other hand, using the NB classifier used in the own pipeline in this work, such an approach does not pose any additional computational challenges.

6.5.2 Model Similarity

Comparing models over time can bring additional insights in the change in the participant behaviour introduced by an intervention. For instance, comparing the model obtained before the change in the participant's diet and after the change can demonstrate whether the intervention was effective. This approach falls under a classic interrupted time series analysis [28] and can aid in performing a Just-In-Time Adaptive Intervention (JITAI) [29] by analysing the effectiveness of the past interventions for a given participant.

The clustering approach used in this work is deemed to be more accurate than the approach used in [12], as it does not require setting a fixed number of clusters in advance. Definition of a maximum allowed distance between the models is seen as a more natural alternative when determining the assignment of participants to behaviour-based groups. The clusters shown in Fig. 6.7 represent the groups of individuals with similar behaviour patterns. Identification of outliers (the participants with uncommon behaviour patterns) can be performed using the same technique, by providing a lower bound of minimum distance signifying an uncommon behaviour.

Finding participants with similar or very different behavioural patterns can be useful in practice for selecting appropriate "buddies" for a participant, such that these participants are put in a group, where the members can chat together about their experiences and reflect on their (similar) behaviours together. The described connectedness of the participants is deemed to be particularly important for a long-term use of an EMA-based system. Note that the participants can be reassigned to different groups over time, since the clustering is done in a data-driven way. Therefore, a perspective presented in Fig. 6.7 can provide a fine-grained insight regarding behaviour similarities between participants.

6.5.3 Model Validation

The purpose of the original study [7], in which the data used in this work was collected, was to predict moments when a participant was about to make an unhealthy eating choice and to provide real-time feedback which could challenge the participant's choice towards a healthier eating alternative. Other predictive models applied to the dataset previously were validated by human experts (psychologists) during and after the study, as described in [12, 13, 21–23]. However, no detailed/complete analysis of individual behaviours was performed in the scope of this work. Therefore, model validity was not checked with human experts and this is still to be done.

Specifically, it is important to verify whether the models and the types of analysis described in this work can be used in an intervention context by providing personalized data-driven feedback in a dedicated EMA study. This means, that prediction quality and model transparency have to be assessed by both the experts conducting the study and the participants of the study receiving the feedback. In this assessment, the qualitative feedback and acceptance also play an important role on par with the quantitative analysis (as presented in this work). Therefore, the next step in validation of the models described in this work is to conduct a dedicated EMA study in which the models are used in a real-life setting and to collect qualitative and quantitative feedback from the experts and study participants.

6.5.4 EMA with Other Features

Additional momentary features collected in an EMA study can be easily incorporated in the described models due to the use of the Bayesian framework. If the assumption of independence between the features holds (Eqs. (6.4) and (6.6) are used for training and prediction, respectively), the prediction performance is not expected to suffer from the curse of dimensionality. If the feature independence assumption is dropped (Eqs. (6.5) and (6.7) are used for training and prediction respectively), the number of possible feature pairs grows exponentially.

For training, this is not a problem, since only the observed pairs are considered/stored. During testing, the issue can be mitigated by the adoption of heuristics, such as, for instance, inclusion of the most frequent N pairs in the calculation of Eq. (6.7). The most frequent pairs can be retrieved from the counts computed using Eq. (6.5) when sorted by value. This way, the common line of the Bayesian framework, in which the infrequent feature values will not have a strong influence on the predictions, is maintained.

For the non-momentary features, meaning the ones which are not expected to alter from one sample to another or change only slightly (for example, the person's age), a different analysis will be required. The impact of the non-momentary features on the results cannot be observed/evaluated using the models described in this work. Therefore, such features are not expected to improve prediction quality.

However, the non-momentary, user-specific features (such as age) can be used when comparing the models of different users. For instance, during the analysis it can turn out that the users whose models are similar, also belong to the same age range. Therefore, the non-momentary features can suggest new insights about the phenomenon analysed in the EMA study.

6.5.5   Personalization of Class Labels

In the EMA study, in which the data used in this work was collected, all the food items were labelled the same for all the participants [7]. Individual dietary restrictions were not considered during the labelling process. One could argue that pasta is not "healthy" given that it is rich of carbohydrates (which in itself is a valid concern, especially if a lot of pasta is consumed regularly), but pasta is a healthier choice than chips, fries, and pizza, for example. Thus, pasta was labelled as a "healthy" choice.

Healthy food for one individual may be unhealthy for another, given their condition (such as allergy). From the data perspective, healthy/unhealthy labelling of food items for individual persons is straightforward. For instance, if someone is allergic to pasta, then pasta is marked as an unhealthy choice for this individual. The model training/evaluation process is not changed, and, in fact, the individual will obtain more suitable feedback, considering own allergy.

However, in practice, it is not always possible to reliably obtain the information about a person's allergies in advance. For example, when the allergy manifests itself over time and was not known to the person before. The same applies to diseases emerging during the data collection process.

Nevertheless, it is indeed important to ask each participant about their allergies, diseases, religious and dietary restrictions before collecting data. This way a more reliable personalisation of the feedback can be provided. Once the onset state of the user has been taken into account, it is also important to track changes in the user's state (for instance, a new disease) during the data collection in order to properly label the healthy/unhealthy choices after each change. This is to be done in the future EMA studies.

6.5.6   Method Limitations

One of the key limitations of the method described in Sect. 6.3.2.1 is its limited adaptability to behavioural changes occurring over time. This is particularly prominent when an established pattern is challenged by a participant over time, as it can be seen in Fig. 6.3 for the lines experiencing abrupt value shifts. As the model's prediction is based on the whole corpus of data available for a particular participant, if a behaviour change occurs, for instance, only on the last day of a week-long monitoring period, the model will not be able to distinguish the recent behaviour change from noise. Only when the new behaviour is observed in a majority of the collected data, the behaviour will be established as an effective participant-specific pattern.

Observing behaviour changes is possible in multiple ways. For example, by performing a type II model comparison (see Sect. 6.3.4.4 for explanation) or by inspecting dynamic behaviour networks (not shown in this work). However, the observed changes are not immediately actionable, and thus the comparison does not directly aid in implementing interventions aimed at behavioural change. To make the short-term changes recognized when applying a model in the intervention context, it is suggested to train the same models, but using different time frames, and then aggregate the predictions made by each of the models. For example, in addition to the model of "overall" behaviour collected for the whole duration of the EMA study, models can be trained using the data collected for the last N days or weeks to provide a desired level of insight granularity. The latter approach is not part of this work and is left for future investigation.

Another deficiency of the method, related to the mentioned limited adaptability, is the unstable prediction performance when two behaviour patterns are (almost) equally probable. This means that either of such behaviours can be taken to predict a class value, which results in class value abruptly shifting from one value to another, especially in early stages of training when little data is available (as can be seen in Figs. 6.3 and 6.4). This shifting can be reduced, for instance, by applying a Moving Average (MA) filter of length k to the class values output by the predictive model. However, no comprehensive solution of this problem is deemed possible.

6.6   Conclusion

The use of the EMA data allows modelling human behaviour in a personalized way. In this work, it was demonstrated that the use of the transparent predictive models for EMA data allows a comprehensive human behaviour analysis. Namely, the analysis was performed at multiple levels by assessing the:

- prediction performance obtained for a given participant compared with a performance of the group;
- model convergence across time;
- development of the behaviour described by a model of one participant compared with the one of another participant;
- clusters of participants with the similar behaviour patterns;
- contributions of individual behaviour components in deciding for a prediction target; and

- correlations between the behaviour components linked to a prediction target of interest.

Moreover, analysis of the models can be performed in a privacy-preserving way, as none of the raw data is shared and the model does not need to contain identifying information. Therefore, in this case transparency does not hinder security.

Although the EMA dataset used in this work is a result of data collection for a period of 8 weeks, the amount of data collected for each participant remains small. Thus, this poses challenges for a cold-start, incremental learning paradigm employed in this work. Therefore, prediction of the human behaviour using the EMA data remains a challenging task. The use of personalization does not always benefit the prediction performance. Moreover, the increased length of history of samples allows decreasing the prediction variability, but it does not guarantee an increased accuracy.

Mitigation of these challenges is possible when the data is collected for a longer period of time or when the data sampling rate is increased. The latter is achievable when using passive monitoring with sensors (for example, Inertial Measurement Unit (IMU) or Global Positioning System (GPS)). Using sensors for data collection and passive monitoring will be considered in the future work.

Acknowledgments We would like to thank Anne Roefs, Bastiaan Boh and Vincent Kerkhofs for their research in scope of the "ThinkSlim" project, which resulted in the EMA dataset used in this paper. The "ThinkSlim" project was supported by NWO. This work is supported by EIT Digital and EIT Health as an Innovation Activity on "Combating Child Obesity". EIT Health and EIT Digital are supported by the European Institute of Innovation and Technology (EIT), a body of the European Union which receives support from the European Union's Horizon 2020 Research and innovation programme.

References

1. Stachl, C, Au, Q, Schoedel, R, Gosling, S. D., Harari, G. M., Buschek, D., Völkel, S. T., Schuwerk, T., Oldemeier, M., Ullmann, T., Hussmann, H., Bischl, B., & Bühner, M. (2020). Predicting personality from patterns of behavior collected with smartphones. *Proceedings of the National Academy of Sciences of the United States of America, 117*(30), 17680–17687. https://doi.org/10.1073/pnas.1920484117
2. Grenard, J. L., Stacy, A. W., Shiffman, S., Baraldi, A. N., MacKinnon, D. P., Lockhart, G., Kisbu-Sakarya, Y., Boyle, S., Beleva, Y., Koprowski, C., & Ames, S. L. (2013). Sweetened drink and snacking cues in adolescents: A study using Ecological Momentary Assessment. *Appetite, 67*:61–73. https://doi.org/10.1016/j.appet.2013.03.016
3. Alvarez, F., Popa, M., Solachidis, V., Hernández-Peñaloza, G., Belmonte-Hernández, A., Asteriadis, S., Vretos, N., Quintana, M., Theodoridis, T., Dotti, D., & Daras, P. (2018). Behavior analysis through multimodal sensing for care of Parkinson's and Alzheimer's patients. *IEEE MultiMedia, 25*(1), 14–25. https://doi.org/10.1109/MMUL.2018.011921232
4. FAO, IFAD, UNICEF, WFP, & WHO (2020). The state of food security and nutrition in the world 2020. transforming food systems for affordable healthy diets. Technical report, FAO, Rome, Italy. https://doi.org/10.4060/ca9692en

5. Bolger, N., Davis, A., & Rafaeli, E. (2003). Diary methods: Capturing life as it is lived. *Annual Review of Psychology, 54*(1), 579–616. https://doi.org/10.1146/annurev.psych.54. 101601.145030

6. Stone, A. A., & Shiffman, S. (1994). Ecological Momentary Assessment (EMA) in behavioral medicine. *Annals of Behavioral Medicine, 16*(3), 199–202. https://doi.org/10.1093/abm/16.3. 199

7. Boh, B., Lemmens, L. H., Jansen, A., Nederkoorn, C., Kerkhofs, V., Spanakis, G., Weiss, G., & Roefs, A. (2016). An Ecological Momentary Intervention for weight loss and healthy eating via smartphone and internet: Study protocol for a Randomized Control Trial. *Trials, 17*(1), 154. https://doi.org/10.1186/s13063-016-1280-x

8. Heron, K. E., & Smyth, J. M. (2010). Ecological Momentary Intervention: Incorporating mobile technology into psychosocial and health behaviour treatments. *British Journal of Health Psychology, 15*(1), 1–39. https://doi.org/10.1348/135910709X466063

9. Atienza, A. A., King, A. C., Oliveira, B. M., Ahn, D. K., & Gardner, C. D. (2008). Using hand-held computer technologies to improve dietary intake. *American Journal of Preventive Medicine, 34*(6), 514–518. https://doi.org/10.1016/j.amepre.2008.01.034

10. Richardson, B., Fuller-Tyszkiewicz, M., O'Donnell, R., Ling, M., & Staiger, P. K. (2017). Regression tree analysis of Ecological Momentary Assessment data. *Health Psychology Review, 11*(3), 235–241. https://doi.org/10.1080/17437199.2017.1343677

11. Kim, H., Lee, S., Lee, S., Hong, S., Kang, H., & Kim, N. (2019). Depression prediction by using Ecological Momentary Assessments, Actiwatch data, and machine learning: Observational study on older adults living alone. *JMIR mHealth and uHealth, 7*(10), e14149. https://doi.org/10.2196/14149

12. Spanakis, G., Weiss, G., Boh, B., Lemmens, L., & Roefs, A. (2017). Machine learning techniques in eating behavior e-coaching. *Personal and Ubiquitous Computing, 21*(4), 645–659. https://doi.org/10.1007/s00779-017-1022-4

13. Spanakis, G., Weiss, G., Boh, B., Kerkhofs, V., & Roefs, A. (2016). Utilizing longitudinal data to build decision trees for profile building and predicting eating behavior. *Procedia Computer Science, 100*, 782–789. https://doi.org/10.1016/j.procs.2016.09.225, international Conference on ENTERprise Information Systems/International Conference on Project MANagement/International Conference on Health and Social Care Information Systems and Technologies, CENTERIS/ProjMAN/HCist 2016

14. Spanakis, G., Weiss, G., & Roefs, A. (2016). Enhancing classification of ecological momentary assessment data using bagging and boosting. In *Proceedings of the 28th IEEE International Conference on Tools with Artificial Intelligence (ICTAI) 2016* (pp. 388–395). Piscataway, NJ: IEEE. https://doi.org/10.1109/ICTAI.2016.0066

15. Lin, X. (2018). Ecological Momentary Assessment (EMA) data: Statistical methods for heterogeneous variance, missing data and latent state classification. PhD thesis, University of Chicago. https://doi.org/10.6082/tkq3-7x06

16. Mikus, A., Hoogendoorn, M., Rocha, A., Gama, J., Ruwaard, J., & Riper, H. (2018). Predicting short term mood developments among depressed patients using adherence and Ecological Momentary Assessment data. *Internet Interventions, 12*, 105–110. https://doi.org/10.1016/j. invent.2017.10.001

17. Sloan, R. H., & Warner, R. (2018). When is an algorithm transparent? Predictive analytics, privacy, and public policy. *IEEE Security & Privacy, 16*(3):18–25. https://doi.org/10.1109/ MSP.2018.2701166

18. Craglia, M., Annoni, A., Benczúr, P., Bertoldi, P., Delipetrev, B., De Prato, G., Feijoo, C., Macias, E., Gómez, E., Iglesias, M., Junklewitz, H., Lopez-Cobo, M., Martens, B., Nascimento, S., Nativi, S., Polvora, A., Sanchez, I., Tolan, S., Tuomi, I., & Vesnić Alujević, L. (2018). *Artificial intelligence: A European perspective.* Publications Office of the European Union. https://doi.org/10.2760/11251

19. Borsboom, D., & Cramer, A. O. J. (2013). Network analysis: An integrative approach to the structure of psychopathology. *Annual Review of Clinical Psychology, 9*(1), 91–121. https://doi.org/10.1146/annurev-clinpsy-050212-185608

20. van der Aalst, W. M. P. (2011). *Process mining: Discovery, conformance and enhancement of business processes.* Berlin: Springer. https://doi.org/10.1007/978-3-642-19345-3

21. Spanakis, G., Weiss, G., Boh, B., & Roefs, A. (2016). Network analysis of Ecological Momentary Assessment data for monitoring and understanding eating behavior. In X. Zheng, D. D. Zeng, H. Chen & S. J. Leischow (Eds.), *Proceedings of the International Conference on Smart Health (ICSH) 2015.* Cham: Springer (pp. 43–54). https://doi.org/10.1007/n-319-29175-8_5

22. Boh, B., Jansen, A., Clijsters, I., Nederkoorn, C., Lemmens, L. H., Spanakis, G., & Roefs, A. (2016). Indulgent thinking? Ecological momentary assessment of overweight and healthy-weight participants' cognitions and emotions. *Behaviour Research and Therapy, 87*, 196–206. https://doi.org/10.1016/j.brat.2016.10.001

23. Roefs, A., Boh, B., Spanakis, G., Nederkoorn, C., Lemmens, L. H. J. M., & Jansen, A. (2019). Food craving in daily life: Comparison of overweight and normal-weight participants with ecological momentary assessment. *Journal of Human Nutrition and Dietetics, 32*(6), 765–774. https://doi.org/10.1111/jhn.12693

24. Iverson, K. E. (1962). *A Programming language.* New York, NY: Wiley.

25. Fisher, R. A. (1936). The use of multiple measurements in taxonomic problems. *Annals of Eugenics, 7*(2), 179–188. https://doi.org/10.1111/j.1469-1809.1936.tb02137.x

26. Traag, V. A., Waltman, L., & van Eck, N. J. (2019). From Louvain to Leiden: Guaranteeing well-connected communities. *Scientific Reports, 9*(1), 5233. https://doi.org/10.1038/s41598-019-41695-z

27. Bastian, M., Heymann, S., & Jacomy, M. (2009). Gephi: An open source software for exploring and manipulating networks. In: *Proceedings of the 3rd International AAAI Conference on Weblogs and Social Media* (pp. 361–362).

28. McDowall, D., McCleary, R., Meidinger, E. E., & Hay, R. A. (1980). *Interrupted time series analysis. No. 21 in quantitative applications in the social sciences.* Thousand Oaks, CA: SAGE Publications. https://doi.org/10.4135/9781412984607

29. Nahum-Shani, I., Smith, S. N., Spring, B. J., Collins, L. M., Witkiewitz, K., Tewari, A., & Murphy, S. A. (2017). Just-in-time adaptive interventions (JITAIs) in mobile health: Key components and design principles for ongoing health behavior support. *Annals of Behavioral Medicine, 52*(6), 446–462. https://doi.org/10.1007/s12160-016-9830-8

Chapter 7
Mitigating the Class Overlap Problem in Discriminative Localization: COVID-19 and Pneumonia Case Study

Edward Verenich, M. G. Sarwar Murshed, Nazar Khan, Alvaro Velasquez, and Faraz Hussain

7.1 Introduction

The Reverse Transcription-Polymerase Chain Reaction (RT-PCR) test quickly became the gold standard for COVID-19 diagnoses [1]. However, the test suffers from two key problems. First, the test can be prohibitively long to administer in areas of dense outbreaks. As an example, the RT-PCR procedure proposed in [38] determines the absence of COVID-19 in approximately 3 h and 50 min. Second, it is particularly prone to false negatives [20]. Furthermore, patients often must be tested several times in order to yield a confident assessment [31]. As a result, clinicians have largely relied on chest X-rays and CT scans for early identification of potential COVID-19 cases based on the manifestations of salient features correlated with positive COVID-19 diagnoses [3, 17, 39].

Naturally, the foregoing has catalyzed a deluge of research in machine learning applications to the COVID-19 pandemic. A review of machine learning applications

E. Verenich (✉)
Clarkson University, Potsdam, NY, USA

Air Force Research Laboratory, Rome, NY, USA
e-mail: verenie@clarkson.edu; edward.verenich.2@us.af.mil

M. G. Sarwar Murshed · F. Hussain
Clarkson University, Potsdam, NY, USA
e-mail: murshem@clarkson.edu; fhussain@clarkson.edu

N. Khan
Department of Computer Science, University of the Punjab, Pakistan
e-mail: nazarkhan@pucit.edu.pk

A. Velasquez
Air Force Research Laboratory, Rome, NY, USA
e-mail: alvaro.velasquez.1@us.af.mil

to resolving COVID-19 problems ranging from the molecular to the societal has been done by Bullock et al. [4]. In particular, there has been significant progress made in the classification and localization of COVID-19 Regions of Interest (RoI) from Computed Tomography (CT) scans, Lung UltraSonography (LUS) imagery, and chest X-rays. However, the lack of COVID-19 data remains a significant challenge to the development of confident predictive assessments from deep learning systems. It has been noted [42] that manifestations such as bilateral involvement, and their spatial regions, can be common in various types of pulmonary conditions i.e. MERS, COVID-19, and SARS, and it is therefore reasonable to train a classifier or localization model on all available types of data. However, this raises the problem of distinguishing COVID-19 from other pneumonia within a single model trained to detect both diseases, particularly given that there are much fewer COVID-19 labels to learn from. These are generally known as the class overlap and class imbalance problems [2, 25] which stem from the similarity in learned features that obfuscate the decision boundary.

Though labeled bounding boxes can be used to train localization classifiers in a supervised manner [21], the scarcity of such data for positive COVID-19 imagery precludes such an approach. Indeed, to the best of the authors' knowledge, existing public datasets do not include such localized labels [8, 9, 35], though some methods in the literature have obtained such labels for other pulmonary conditions with the help of radiologists [44]. At any rate, the procurement of such labels can be labor-intensive, and it is desirable to autonomously localize RoIs without them. To that end, we exploit the discriminative localization proposed by Zhou et al. [43] which does not require localized labels, but rather exploits the Class Activation Maps (CAMs) computed at the final convolutional layer for each class in a CNN model. Given a class of interest, such as COVID-19, the salient regions in the COVID-19 CAMs can then be upsampled to the size of the original image in order to localize the features most conducive to the classification of the chosen class.

As one of the first tests to be performed for patients suspected to have SARS, MERS, or COVID-19, chest radiographs (X-rays) can identify various pulmonary abnormalities, but as reported in [42], spatial locations of these manifestations are also important, and can help the subject matter expert in distinguishing between similar infections. However, spatial annotations of chest radiographs rarely persisted in patient data, often because further, more sensitive tests are administered to confirm infections. This results in potentially useful chest radiograph data in the form of image-level labels that can be used to train convolutional neural networks that classify chest radiographs as exhibiting certain infections. More specifically, in the case of disease classification, an image-level label is a chest radiograph of a patient that was confirmed to have the disease at the time the X-ray was taken, but spatial locations of associated abnormalities are not available for various reasons. Figure 7.1 further explains the differences between image and object-level labels.

An example of a large collection of chest radiograph image-level labels is the National Institute of Health dataset [37] comprised over 100,000 frontal view chest X-rays. The dataset is accompanied with text labels of disease findings that were mined from patient records for each image. We can use this data to train classifiers

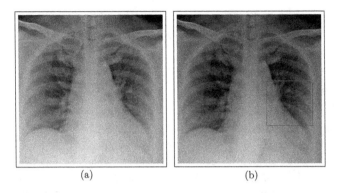

(a)   (b)

Fig. 7.1 Image (**a**) represents a chest radiograph (X-ray) representing an image-level label. A medical finding accompanies this image in the form of a text label stating that COVID-19 was identified in this X-ray, however spatial location of associated manifestations is not specified. Image (**b**) constitutes an object-level label where a finding of COVID-19 is also accompanied with spatial localization of associated manifestations in the form of a red bounding box. A model trained to identify spatial locations of disease associated manifestations can significantly improve explainability of its predictions. (**a**) Image-level. (**b**) Object-level

of diseases, but the lack of spatial locations of manifestations associated with these diseases makes it challenging to interpret the output of such classifiers. In order to increase the confidence of such predictions in the absence of spatial localizations in our training data, we would like to employ weakly supervised localization methods that do not require spatial localizations, or object-level labels, for training. The main idea of using weakly supervised localization for increasing confidence in predictions produced by image classifiers is to identify spatial locations within the image which are used to provide a given prediction.

This leads us to the main problem associated with real world data, which is noise, class overlap, and class imbalance. The main challenge to interpreting deep learning predictions is the aleatoric uncertainty that arises from neuron co-adaptation when the model is trained on labels with highly overlapped classes. This significantly degrades the quality of weakly supervised localization, when using image-level labels, thus making it difficult to assess whether the model is concentrating its attention to relevant regions of the image specific to some target class. We show in this paper that performing weakly supervised localization using state-of-the-art approaches suffers from spurious localization in the case of gradient based approaches, and highly broad and uncertain localization in the case of standard class activation maps. This uncertainty makes it difficult to explain the predictions that the model makes to the subject matter expert.

The main contribution of this work is a *novel method that mitigates the neuron co-adaptation effect* that arises in multi-class classifiers trained on image-level data with significant class overlap. We show that *our method successfully reduces aleatoric uncertainty when interpreting output of multi-class models* as a means of performing weakly supervised localization of areas of interest. Our technique works by *exploiting current fundamental architectures of convolutional neural networks*, hence making the approach applicable to many existing applications.

7.2   Related Work

Given the visual salient cues and format of CT, LUS, and X-ray imagery, there have been various deep learning classifiers leveraging Convolutional Neural Networks (CNNs) applied to the medical diagnosis of pneumonia [28] and COVID-19 [23, 40, 41], among other diseases. In the case of COVID-19 detection, though these classifiers can achieve very high precision rates, such classifiers are often moot when it comes to quantifying decision uncertainty. In [13], an approach is proposed to address this based on Bayesian CNNs to estimate the degree of decision uncertainty in existing COVID-19 deep learning classifiers.

Much of the proposed work on COVID-19 classification using CNNs doubles as COVID-19 RoI localization. This is due to the popularity of segmentation-based approaches to classification through the use of UNets. In [16], a UNet architecture is trained using approximately 22000 pneumonia labels in order to localize regions of interest associated with pneumonia. Similarly, [29] leverages a pre-trained UNet in order to obtain the masks for lung region segmentation. This was concatenated to the original CT volume and used as input to a 3D CNN used for classification. Similar approaches using UNet segmentation have also been explored in [6, 7, 14, 15, 19, 26].

UNet has also been extended with attention gates to increase the accuracy of COVID detection. In [30], spatial transformers networks are used to localize COVID-19 RoIs in LUS images. An attention mechanism within CNN architectures was also proposed in [5] to classify RoIs returned by a 3D CNN trained on CT volumes. These CT scans were pre-processed to extract the relevant pulmonary RoIs. In [7], a nested UNet architecture is proposed to classify and localize COVID-19. This architecture, known as UNet++ [44], was trained on a dataset of CT scans with RoIs labeled by radiologists in order to yield accurate localizations. A similar approach using a pre-trained Inception network is proposed in [36].

The main difficulty in applying these state-of-the-art approaches to novel situations, such as the COVID-19 pandemic, is that training labels or images with annotated spatial regions of interest, are often not available and are often prohibitively expensive to obtain in time. Utilizing only image-level labels in the chest radiograph domain, which typically become available sooner due to the prevalence of the chest X-ray exam [42] to screen any suspected pulmonary infection, still allows image classification to be effectively trained. However, a model prediction, even if accompanied with a probability of its prediction, does not communicate relevant information to the subject matter expert in terms of where in the image did the model base its prediction on. Weakly supervised localization is a way to mitigate the problem of missing object-level labels by producing prediction explanations in the form of RoI in models that were only trained on image-level labels. However, as we show in this paper, even state-of-the-art weakly supervised methods can suffer from spurious correlations within training data, potentially explaining their prediction with irrelevant data. In this paper we explore two main strategies for weakly supervised localization, gradient based pixel attributions first proposed in [32] and Class Activation Mappings (CAM) proposed in [43], with the latter approach being the basis of our method.

CAM-based approaches have already been explored to aid in the localization of COVID-19 [33]. Our prior work [34] introduced Amplified Directed Divergence (ADD), which used an ensemble of models to localize COVID-19 RoIs by training two binary classifiers for detecting viral pneumonia and COVID-19, respectively. A directed divergence kernel is then employed to compute the difference in CAMs between the two classes. This work describes a generalization of ADD, which is significant in two key ways. First, we leverage CAM-based localization of overlapping and unbalanced classes within a single model. This is beneficial because COVID-19 data is limited and since viral pneumonia shares some of the underlying features associated with a COVID-19 diagnosis, it makes sense to train a single classifier on both types of data. This also makes our approach easily implementable within most existing COVID-19 classifiers. Second, our approach enables weakly supervised localization in models that were trained on labels where any number of overlapped classes can be present. We show that our new approach performs better weakly supervised localization than gradient based methods and standard class activation based approaches by using a publicly available image classification model pre-trained on ImageNet labels, all without fine-tuning nor altering its architecture.

7.3 Discriminative Localization

In this section we describe discriminative localization approaches, also referred to as weakly supervised object localization, because the convolutional neural networks utilized in these approaches are typically trained for image classification tasks on *image-level labels only*, with no additional spatial information such as object-level bounding boxes. As previously stated, we emphasize these approaches because the object-level data necessary for state-of-the-art object detection models is often not available, while ground truth for image-level data can be collected indirectly.

In earlier work [34], we described a scenario when image-level labels of COVID-19 X-ray samples can be acquired, without the costly labeling process by radiologists. So, X-ray samples of patients confirmed to have COVID-19 through RT-PCR testing are collected. In that scenario, it is known that a given sample (image-level label) contains features related to COVID-19, but there is no information on the spatial regions of those features (object-level labels).

In this section, we first describe two current approaches to discriminative localization, viz. Class Activation Maps and Saliency Maps, with the former being the basis for our approach to dealing with discriminative localization in the class overlap setting. Next, we describe our approach called *Amplified Directed*

Divergence (ADD) that works with ensembles of models to deal with class overlap and class imbalance [34]. We then introduce *Scaled Directed Divergence (SDD)* that generalizes *ADD* to multi-class models.

In order to intuitively describe the presented methods, some examples in this work use natural imagery from the ImageNet dataset [10]. This dataset consists of over 14 million images that map to 1000 classes, such as cat, tree, and sports car among others. The first state-of-the-art result in image classification using deep neural networks was achieved using a model called *AlexNet* [18], which was trained using this dataset. It has since become a standard dataset for training image classifiers in the natural imagery domain, with many publicly available models fully pre-trained on the entire dataset. In order to illustrate class overlap in the natural image domain, we utilize the following three ImageNet classes as defined in the ImageNet 1000 class index:

- 479: "car wheel"
- 751: "racer, race car, racing car"
- 817: "sports car"

The "car wheel" class represents our *target class* that has significant features overlap with the other two classes (751: race car and 817: sports car). The number preceding the name of a class is its class index within ImageNet. When using publicly available images of passenger cars, to perform image classification with models pre-trained on ImageNet, both classes 751 and 817 are frequently returned and are typically found in the top-3 class results together, meaning these classes are very similar. We include both classes in our experiments to increase the level of overlap with our target class (car wheel: 479), and *may refer to classes 751 and 817 as* car *interchangeably*. Furthermore, by utilizing these classes we simplify reproducibility of our results, as most widely used neural network training libraries provide implementations of many popular convolutional neural network architectures pre-trained on ImageNet.

7.3.1 Class Activation Maps

At the heart of our proposed method is the use of class activation mappings proposed by Zhou et al [43]. The key to performing discriminative localization with this approach is for a given convolutional network architecture to have a global average pooling (GAP) layer that outputs the *spatial average* of each feature map at each unit of the last convolutional layer. These spatial averages are then used as inputs to a fully connected layer to perform classification. By projecting weights associated with a given predicted class from the final connected layer to convolutional feature maps, class contributing regions of an image are identified. We believe that the use of global average pooling instead of global max pooling, as in [24], is key to avoid focusing on spurious spatial regions of an image with respect to the predicted class.

Formally, for a given input image, let $f_k(x, y)$ be the activation of filter k at the last convolutional layer, where (x, y) is the spatial location. Then, for each k, the global average pooling (GAP) layer outputs F_k, defined as $\sum_{x,y} f_k(x, y)$, which are then used as input features into the final connected layer, whose output is $\sum_k w_k^c F_k$, where w_k^c is the weight for class c and filter k. Then, the class activation map M_c is given by

$$M_c(x, y) = \sum_k w_k^c f_k(x, y) \tag{7.1}$$

where, for each class c, the number of weights w equals to the number of filters k. Figure 7.2 shows a visual representation of projecting weights from the final fully connected layer to compute the class activation map.

7.3.2 Saliency Maps with Backpropagation

Simonyan et al. [32] proposed a method for querying a convolutional network classifier about the spatial support of a given class for a given image. Specifically, given an image I_0 and a class c, and a model with a given class score function $S_c(I)$, they rank the pixels of I_0 based on their contribution to the class score $S_c(I_0)$. This is done by computing the derivative vector w of a scoring function S_c with respect to image I at an image location I_0. The saliency map $M \in R^{m \times n}$, where m and n corresponds to rows and columns of the input image, is then obtained by rearranging the elements of w. For multi-channel images, a saliency map is computed for each channel. To compute a single map for all channels, a maximum magnitude of w is taken across all color channels. The intuition here is that the magnitude of gradients w indicates which pixels can be changed the least to affect the class score the most, thereby corresponding to a localized target object within the image as shown in Fig. 7.3.

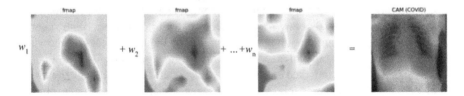

Fig. 7.2 Class activation map (CAM) for a given class c is computed by performing a weighted sum of class weights w_k^c with feature activations of units k residing in the last convolutional layer of a Convolutional Neural Network. For example, in a ResNet 152 architecture, the last convolutional layer contains 2048 units with a mapping resolution of 7×7. Then each unit in the final fully connected layer takes as input 2048 inputs, where weights w_k^c of the final layer indicate the importance of Global Average Pooling output F_k to class c. Hence, we multiply each 7×7 feature map by a scalar w_k^c and sum them to produce a class activation map

Fig. 7.3 Saliency maps computed on a sample image using a ResNet152 model pre-trained on ImageNet. *The leftmost* image represents our input image, which was classified by the network as class 817: "sports car, sport car," as per ImageNet 1000 class ID list. The *second* image from the left represents gradients across three RGB channels. The *hhird* image shows a saliency map computed by taking maximum gradients across all channels. The *fourth* image overlays the max gradient saliency map on the original image, showing the spatial support for features that localize the target object within the input image

The computation of an image-specific saliency map for a given class only requires a single backpropagation pass, hence it is efficient. This is different from another method used for visualizing learned representations of classes, where an input image with pixel intensities initialized to zero is optimized using backpropagation while holding filter weights of the model constant to what was learned during training [11]. The result is a numerically computed image that represents a visual class representation learned from an entire class of images, hence it is not image-specific. There is however a potential utility to such representations in the image-specific setting. Geirhos et al. [12] discuss apparent bias of trained convolutional neural networks toward texture versus shape, hence it may be possible to generate images representing textures that are learned by image classifiers and observing their presence in specific image samples.

Although pixel-level backpropagation methods have shown discriminative localization ability, there are drawbacks that can be shown empirically. Specifically, the method is prone to focusing attention on spurious features of an image, even when the network returns a correct classification. In fact, a CAM-based method that utilizes Global Average Pooling [43] has been shown to outperform both saliency maps [32] and Global Maximum Pooling [24] methods in weakly supervised localization. As an example of errors due to the focus on spurious features, Fig. 7.4 shows saliency maps computed on another sample input image. Although the model classified the image as a car (class 817), the saliency map is focused on a spurious feature, apparently a road or parking sign. The same could be seen in the radiology domain, where saliency maps focus on X-ray markings.

The CAM-based approach with global average pooling avoids this problem due to all filters contributing to the attention region in a weighted manner, instead of taking only maximum values as in both saliency maps and global max pooling approaches. Figure 7.5 shows class activation maps computed for our three classes of interest (i.e. 479, 817, 751), where spurious features that were focused on by saliency maps did not appear in the CAMs. However, significant overlap between

Fig. 7.4 Saliency maps focused on spurious features. The *left* image represents our input image, classified correctly as class 817 (car). The *second* image from left represents gradients across three RGB channels. The *third* image from the left shows a saliency map computed by taking maximum gradients across all channels. The *rightmost* image overlays the max gradient saliency map on the original image, showing the spatial support for features that localize the target object within the input image

Fig. 7.5 The same input image as in Fig. 7.4 and Class Activation Maps are computed for three ImageNet classes of interest, carwheel, race car, and sports car. This method is not prone to result in attention on spurious features. However, as can be seen from the three leftmost images, significant overlap exists between the classes for which CAMs were computed

the CAMs exists, specifically between class 479 (car wheel) and both 817 (sports car) and 751 (race car).

7.3.3 Amplified Directed Divergence with Ensembles

Figure 7.5 illustrates the difficulty of performing weakly supervised localization when class overlap is present within our data. The class activation maps generated for the first image, class 479 (car wheel), and second image, class 817 (sports car) are similar and significantly overlap each other in terms of their spatial attributes. We note that the convolutional neural network used to generate these class activation maps was fully trained on the ImageNet database for image-level classification. Without localized object-level labels, there is a higher level of uncertainty regarding where the model focused its attention to identify its target class. This is referred to as *aleatoric* predictive uncertainty [13] and is often increased by the presence of overlapping classes within data. The difficulty with aleatoric uncertainty is that it cannot be easily reduced with more training data, unlike *epistemic* uncertainty which

Fig. 7.6 Example of Amplified Directed Divergence kernel applied to class activation maps computed from two expert models. The first image shows the class activation map computed on a car wheel classification from an expert model trained to classify car wheels. The second image shows a class activation map on a car classification from an expert model for cars. The third image shows a class activation map that better localizes the target class (car wheel) computed using the kernel method

is associated with model parameters and thus can be reduced with more training samples.

To mitigate aleatoric uncertainty, we had [34] introduced Amplified Directed Divergence, a kernel function that accepts two class activation maps from expert models, each trained on specific overlapped classes, and extracts activations that are relevant to one of them, i.e. target class (car wheel) as seen in Fig. 7.6. The result is a new class activation map that better localizes objects of interest in the presence of class overlap. Formally, let $x \in R^{mxn}$ be the first class activation map calculated using the output of the first expert model M, and let $x' \in R^{mxn}$ be a second class activation map calculated using the output of a second expert model M'. Then, the new class activation map showing directed divergence of overlapped classes is given by

$$ADD(x, x') = e^{\alpha(x/max(x) - x'/max(x'))} \tag{7.2}$$

where parameter α controls the amplification of the directed divergence of the two class activation maps. Increasing the value of α results in more focused regions, as shown in Fig. 7.7. We also note that normalization of class activation maps is necessary because they are calculated from two separate models that are trained independently, which can result in different extremes in magnitudes of activations when fed the same image. Additionally, any normalization method of the class activation map tensors can be employed, so long as it is consistent with both tensors. Finally, since Amplified Directed Divergence is a directed method, $ADD(a, b) \neq ADD(b, a)$.

Figure 7.8 shows a high-level architecture of the Amplified Directed Divergence method that utilizes expert network ensembles to mitigate the class overlap problem. The use of two networks has advantages and disadvantages as compared to utilizing a single multi-class model, as we will explore next. One advantage is that this

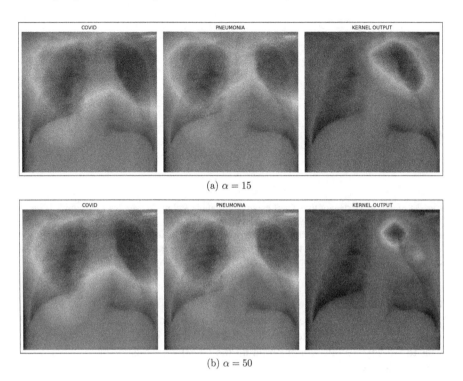

(a) $\alpha = 15$

(b) $\alpha = 50$

Fig. 7.7 Role of the amplification parameter in Eq. 7.2. As the amplification parameter α is increased, the COVID-19 localization in the heat map output by the kernel function becomes more concentrated. More specifically, by increasing amplification of directed differences in activation maps, small differences contribute less to the CAM computed using Eq. 7.2, making spatial localization heatmaps concentrate on the highest activations associated with the target class. Decreasing α results in more spread out heatmaps accounting for smaller differences in class activation maps. This parameter can also be used to adjust the dimensions of a computed bounding box when using class activation maps for weak labeling, i.e. computing locations of bounding boxes based on class activation maps. (**a**) $\alpha = 15$. (**b**) $\alpha = 50$

architecture allows us to use models that were trained for different purposes, i.e. classification using image-level data and object localization using more expensive object-level data, as long as they both include Global Average Pooling of final convolutions with the same mapping resolution in the down-sampling stage of their architecture.

Consider for example a network trained to localize or segment specific lung regions on expertly annotated X-ray images of Pneumonia cases. These types of images could be more readily available due to prevalence and length of time; such cases have been treated and studied. On the other hand, when a novel condition emerges, such as the COVID-19 pandemic, expertly annotated X-rays may not be available for training a model with object-level labels. We can however train COVID-19 classifiers [25] using image-level labels, since labels for such cases can often be obtained indirectly through collection of X-ray imagery of confirmed

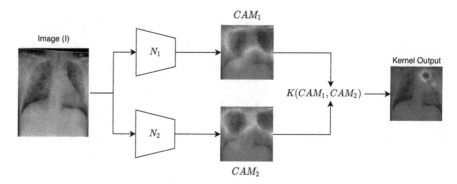

Fig. 7.8 High-level view of the expert network ensemble architecture as described in [34]. Given a target class c_1 and a second possibly overlapping class c_2, two binary expert models N_1 and N_2 are used to calculate class activation maps CAM_1 and CAM_2 for their respective classes. Class activation maps are then passed to the Amplified Directed Divergence kernel to compute the final class activation map that more narrowly localizes a spatial region associated with CAM_1

COVID-19 patients where their diagnoses were confirmed through other methods, e.g. RT-PCR test [1]. The two models can now be used with the Amplified Directed Divergence architecture to extract localized regions that are more relevant to COVID-19, all without training the COVID-19 model for localization or segmentation.

A *potential weakness* of this approach is that it is difficult to utilize it for instances when multiple possible classes exhibit significant features overlap with our target class. Consider the example in Fig. 7.5, where our target class is *car wheel* (479), while classes 751 and 817 also produce class activation maps that are both similar to our target class. This behavior is an artifact of neuron co-adaptation, as samples of each class can influence the discriminative function of all other classes through shared architecture. In cases where we need to perform discriminative localization with multiple overlapping classes, the method needs to be adapted to the multi-class setting.

7.3.4 Scaled Directed Divergence (SDD)

To address the multiple overlapped classes problem, in the multi-class model setting, described above (also see Fig. 7.5), along with the problem of focusing on spurious features shown in Fig. 7.4, we present our generalization of Amplified Directed Divergence to the multi-class model setting. Formally, let $S = \{x_0, \ldots, x_i\}$ be the set of Class Activation Maps corresponding to $i + 1$ classes of a multi-class model. Let $T(t) = S \setminus \{x_t\}$ be the set of Class Activation Maps where the target class CAM x_t is removed from set S. Then, a new CAM for a given target class is defined as

Fig. 7.9 Discriminative localization through class activation maps in class overlapped setting using a multi-class model pre-trained on the full 1000 class ImageNet dataset. The first image (from the left) shows the class activation map computed for the car wheel (479) class. The second and third images show class activation maps for classes 817 (sport car) and 751 (race car). The fourth image shows the class activation map computing using Scaled Directed Divergence (479K) for the car wheel class, showing much better localization without the use of object-level labels

$$CAM_{SDD} = e^{\alpha(x_t|S| - \sum_{k \in T(t)} x_k)} \tag{7.3}$$

where parameter α amplifies or decreases divergence values to be exponentiated, while $|S|$ scales activations of the original target class activation map by the cardinality of S. We note that the set of class activation maps S does not have to be exhaustive of all the classes learned by the model, but must contain our target class and classes with possible overlap. For specialized models (e.g. COVID-19, Viral Pneumonia, No Finding), it is simpler to include all classes. Figure 7.9 shows our method performing discriminative localization on the car wheel (479) class using a ResNet152 model pre-trained on the full 1000 class ImageNet dataset. We note that the model was not fine-tuned by us in any way, the much improved localization in the image (479K) is the result of applying our method to the three class activation maps. In this example, set S consisted of class activation maps for classes (479, 751, 817) and x_t was the class activation map for class 479.

The effect of neuron co-adaptation that arises from training samples with overlapped features is significantly reduced, while allowing similarities in training data to be utilized for mitigating class data imbalance and lack of class data in general.

7.4 Experiments

In this section, we describe our experiments in applying our approach for weakly supervised localization in two domains. First, we experiment using publicly available image classifiers that were pre-trained on the full ImageNet dataset. The purpose of this set of experiments is to empirically validate our approach using visually intuitive data. Second, we apply our approach to our own multi-class (COVID-19, Viral Pneumonia, No Finding) models that were trained on the Kaggle COVID-19 dataset. Due to the novelty of COVID-19, to the best of our knowledge and at the time of this work, there were no publicly available annotations of X-rays representing COVID-19 related (RoI) regions of interest. Therefore, our empirical validation of our approach using natural imagery serves as the next best measure of effectiveness. As our base architecture, we utilize ResNet152 convolutional neural network pre-trained on the full ImageNet dataset. For experiments using natural imagery, no additional training or architecture modifications were performed. For the X-ray domain, we used the pre-trained ResNet152 model as a feature extractor and fine-tuned it using the Kaggle COVID-19 dataset. We also note, that any convolutional neural network architecture can be adapted to this approach by adding a Global Average Pooling layer immediately after the last convolutional layer of the network.

7.4.1   Method

Prior research has shown [43] that increasing the mapping resolution of the final convolutional layer increased the discriminative localization performance of a network. For pre-trained models, this is typically done by removing one or more convolutional layers at the end of the network and adding a new layer with appropriate kernel size, stride, and the number of units to increase the final mapping resolution. For this work, we forgo these modifications and apply *SDD* to existing implementations of ResNet152, as our goal is to verify the effectiveness of our method regardless of final mapping resolution. In our case, the final mapping resolution of the last convolutional layer is 7×7. This means that the dimensions of computed class activation maps will also be 7×7, and given the shape of our input images (224×224), the resulting up-scaling factor required to overlay class activation maps over original images is 32. To do this, we utilize a computer vision technique called *bilinear upsampling*, where each pixel from the computed CAM is moved in a given direction using some scaling constant, and pixels with missing intensity values resulting from this movement are computed by taking the weighted average of four diagonal pixels surrounding the pixel being computed.

In order to compute class activation maps, filter activations must be captured from the last convolutional layer during inference, i.e. passing an image through the network in order to classify it. Methods for capturing filter activations depend on the tools used to implement the network, which in our case is the PyTorch library. The specific ResNet152 layers that were used in our computations of class activation maps were the following:

- **model.layer4[3].relu** is the last convolutional layer before the Global Average Pooling layer. Note that **layer4** refers to the fourth convolutional stage of ResNet architectures. We also extract activation values after the ReLU layers to remove any negative noise.
- **model.fc.weights** is the location of class weights. Note that before projecting weights back to activations, class specific weights must be specified.

As a model is loaded, a hook is attached to the last convolutional layer before the network is used for inference. Once an image is fed to the network as input, activations of that layer are saved for that image and class specific weights from the last layer are then used to compute class activation maps for each class of interest. These class activation maps are then passed to the Scaled Directed Divergence kernel function to compute the new class activation map. All of the computed class activation maps are upscaled to the size of the original image using bilinear upsampling and are overlaid over the original image.

7.4.2 COVID-19 and Pneumonia Data

Our multi-class X-ray classifier was trained using publicly available COVID-19 radiology data [27]. As mentioned earlier, only image-level labels were available for this data, meaning no localization metadata of regions of interest are present. The dataset was randomly split into 60/20/20 train, validate, test partitions as shown in Table 7.1. Of the three classes within this dataset, COVID-19 samples contain roughly six times fewer samples than viral pneumonia or samples with no findings, i.e. Normal. During network fine-tuning and inference stages, samples are resized to 224×224 pixels and are normalized using per channel mean [0.485, 0.456, 0.406], and standard deviation [0.229, 0.224, 0.225] computed on the ImageNet dataset. This was done because our custom radiology classifier uses an ImageNet pre-trained ResNet152 as a feature extractor and backbone that was fine-tuned to the new dataset.

7.4.3 COVID-19 AND Pneumonia Classifier

We now describe the convolutional neural network used in the application of our method to improve weakly supervised localization in the X-ray domain. We trained a multi-class image classifier on image-level labels of X-ray images in the COVID-19 dataset. We have seen that propagation of filter activation values is responsible for classification. It then follows that projecting the learned class weights back to filter activations in order to compute class activation maps improves the quality of localization as model accuracy improves (Fig. 7.10).

Table 7.1 Data partitions of the COVID-19 Radiology Dataset [27]

Partition	COVID-19	Normal	Viral pneumonia
Train	131	804	807
Validate	44	269	269
Test	44	268	269

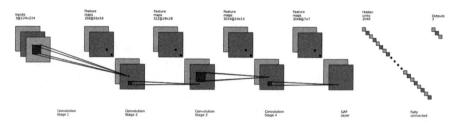

Fig. 7.10 Schematic of our X-ray image classifier based on the ResNet152 architecture. Image input size is 224 × 224, with four convolutional stages common to variants of the ResNet architecture. Output mapping resolutions of the four convolutional stages are as follows: 56x56 at stage 1, 28 × 28 at stage 2, 14 × 14 at stage 3, and 7 × 7 at stage 4, the final convolutional stage. The final connected layer has 3 outputs compared to the original 1000

Fig. 7.11 Validation accuracy progress for AdamW and SGD optimized models during fine-tuning for 30 epochs

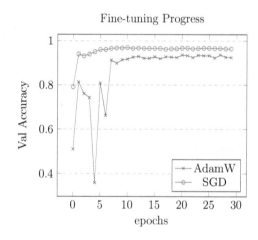

As part of our model selection process, we trained two classifiers using different optimization methods, *AdamW* [22] and *Stochastic Gradient Descent*. This was done by replacing the final layer of a pre-trained ResNet152 model to reflect our new classes, then fine-tuning each model for 30 epochs on the X-ray dataset. Replacing the final layer with reduced numbers of units has a beneficial effect on training speed, as the original ResNet152 model contains 2,048,000 trainable parameters just in the final layer, which we reduced to just 6144, significantly reducing epoch times. We checked network performance after every epoch using the validation set, and kept track of the best performing set of weights. After 30 epochs, the best set of weights based on the validation set performance persisted as the final trained model. We then tested both models against the test set and selected the one with higher accuracy as our multi-class classifier. Figure 7.11 shows validation accuracy progress while fine-tuning both models. Table 7.2 shows a confusion matrix for each of the two models as evaluated on the test set, while Table 7.3 shows model performance metrics.

Table 7.2 Confusion matrix for SGD and AdamW optimized models on the test set of the COVID-19 Radiology dataset [27]. Class designations are as follows: COV Covid-19, NOR no findings, and PNE viral pneumonia

		Predicted SGD					Predicted $AdamW$		
		COV	NOR	PNE			COV	NOR	PNE
True	COV	(44)	(3)	(0)	True	COV	(43)	(3)	(4)
	NOR	(0)	(263)	(30)		NOR	(1)	(258)	(32)
	PNE	(0)	(2)	(239)		PNE	(0)	(7)	(233)

Table 7.3 Performance metrics for models trained using SGD and AdamW optimization techniques. Both classifiers are based on the ResNet152 architecture pre-trained on the entire ImageNet 1000 dataset and fine-tuned on the COVID-19 Radiology dataset

	Accuracy	Sensitivity	Precision	F1 Score
ResNet SGD	**0.94**	**0.957**	**0.942**	**0.969**
ResNet AdamW	0.919	0.942	0.906	0.958

Bold values indicate better performance of the top row (model)

7.4.4 Scaled Directed Divergence with Natural Imagery

In this section, we report our weakly supervised localization results on a publicly available image classifier pre-trained on the ImageNet 1000 dataset. In addition, we report results of applying saliency maps, a gradient based technique for weakly supervised localization, to the same set of images.

Figure 7.12 shows rows of images that overlay class activation maps computed for three overlapping classes, car wheel (479), sports car (817), and race car (751), with the last image showing a new class activation map computed using our SDD method for our target class (479). Our method localizes our target class even in the presence of significant overlap with other classes and without any object-level labels. We note again, that no additional training or fine-tuning was performed on the pre-trained ResNet152 model. In addition, our method does not focus on spurious regions as has been observed using the gradient method on the same imagery.

Figure 7.13 shows our results of applying a gradient based weakly supervised localization method, saliency maps [32], to the same model using the same set of images. We observe that the method performs well in localizing some images, but also focuses on spurious regions in other images. In the first row, a road or parking sign is picked up, while in the fourth row a marking at the top of the image is focused on, even though in all cases the model returned the same classification.

As stated, the purpose of these experiments is to provide empirical validation of our method using a visually intuitive domain. Given the absence of object-level or regions of interest labels for currently available COVID-19 X-ray data, we hope that the shown performance of our method on the ImageNet natural imagery dataset provides some intuition on its behavior when applied to COVID-19 data.

Fig. 7.12 Results of applying Scaled Directed Divergence to a pre-trained ResNet152 multi-class image classifier using random car images from Google Search. First three images in each row show class activation maps computed for classes 479 (car wheel), 817 (sports car), and 751 (race car), respectively. The fourth image shows better car wheel (479) localization using our method

Fig. 7.13 Results of applying saliency maps [32], a gradient based approach to weakly supervised localization. The method was applied to the same set of images as our SDD method in Fig. 7.12. Note that the first and fourth rows of images illustrate potential problems with the gradient approach associated with focusing on spurious regions in images

7.4.5 Scaled Directed Divergence Applied to Chest X-rays

We now present a scenario of utilizing Scaled Directed Divergence as a diagnostic aid when applied to pulmonary disease classification. Specifically, our aim is to explain predictions made by our disease classification model by providing spatial regions within the X-ray image that the model based its prediction on. As stated earlier, our multi-class pulmonary disease model was trained to classify X-ray images as one of three classes: COVID-19, viral pneumonia (PNEUM), and no finding (NORMAL). Model training was done using only image-level labels, meaning no spatial annotations of disease manifestations were available. We then used samples from our test set with known labels to classify them using the model and computed class activation maps for each class. We note that regardless of the actual predicted class, CAMs can be computed for each class represented in the model for every image. The intuition here is that given an image sample, it is possible that the model made a close call between COVID-19 and pneumonia, or that in fact both conditions were present with highly overlapped manifestations. This can result in highly overlapped class activation maps for both diseases or very large class activation maps that cover entire lung regions, providing little information regarding disease manifestations in specific regions. Finally, computed class activation maps of all classes were passed to the Scaled Directed Divergence kernel function, described in Sect. 7.3.4 Eq. (7.3), specifying our target class, in order to compute a new class activation map that localizes regions within the image that explain target class predictions from the classification model. Figures 7.14 and 7.15 show the results of this process. In Fig. 7.14 our target class is COVID-19, and Fig. 7.15 uses viral pneumonia (PNEUM) as the target class. More specifically, the fourth image in each row is labeled with a target class appended with a letter K to signify kernel method output.

Next, we describe our results on applying Scaled Directed Divergence to weakly supervised localization of regions of interest in COVID-19 and Viral Pneumonia imagery in more details. Figure 7.14 shows class activation maps computed for three classes learned by our COVID-19, NORMAL, PNEUMONIA (Viral Pneumonia) image classification model when COVID-19 confirmed images are provided to the network. In this case, the model classifies all images as COVID-19, which in this case is our target class for which we would like to compute a new class activation map that better localizes the region that the model thinks is responsible for the COVID-19 classification. For each image row consisting of five images in Fig. 7.14 we have the following from left to right: image 1 shows the class activation map for class COVID-19, image 2 shows the CAM for class NORMAL, image 3 shows the CAM for the viral pneumonia (PNEUM) class, and image 4 shows the output of our SDD (Sect. 7.3.4 Eq. (7.3)) kernel function that localizes and explains the prediction of the *target class*, in this case COVID. The last image in each row is the original image passed to the model, where each of the CAMs in the first four images is superimposed over the original image.

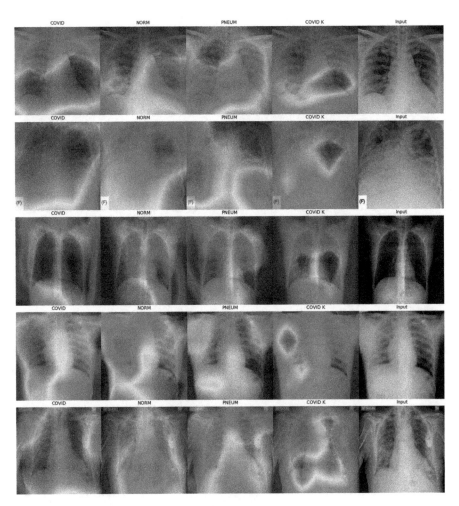

Fig. 7.14 Results of applying our Scaled Directed Divergence method to our custom ResNet152 multi-class image classifier on X-ray images classified as COVID-19 by the model. For each row, the first three images contain class activation maps computed for classes COVID-19, NORMAL, and Pneumonia, respectively. Image four contains a class activation map computed with our method that localizes regions specific to our COVID-19 target class. The last image is the original input image

Figure 7.15 shows the same process applied to images that were classified as Viral Pneumonia (PNEUMONIA) by our model. In this case, our target class for which we wish to compute a new class activation map using Scaled Directed Divergence is Viral Pneumonia. As before, the fourth image in each row represents the class activation map computed using our method and labeled with the predicted class appended with a letter K.

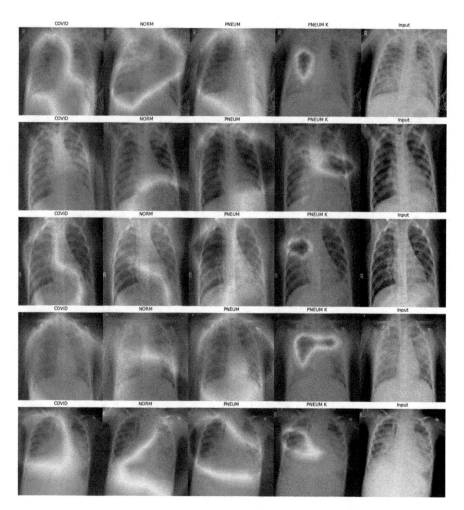

Fig. 7.15 Results of applying our Scaled Directed Divergence method to our fine-tuned ResNet152 multi-class image classifier on X-ray images classified as Viral Pneumonia by the model. For each row, first three images contain class activation maps computed for classes COVID-19, NORMAL, and Viral Pneumonia respectively. Image four (PNEUM K) contains a class activation map computed with our method that localizes regions specific to our PNEUMONIA target class. The last image is the original input image

Finally, Fig. 7.16 shows a sample image classified as Viral Pneumonia, where the first row of images shows weakly supervised localization using the gradient method, i.e. Saliency Maps, while the second row shows weakly supervised localization using Scaled Directed Divergence. The purpose of this example is to illustrate the tendency of the gradient method to focus on spurious regions, even during correct classification. As can be seen in the example, X-ray markings at the top left corner of the image are considered as part of the region making the main contribution to

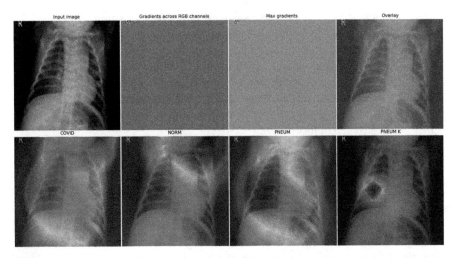

Fig. 7.16 Results of applying Saliency Maps [32] (top row) and Scaled Directed Divergence (second row) to our fine-tuned ResNet152 multi-class image classifier on a sample of X-ray images classified as Viral Pneumonia by the model. In the *top row*, the first image shows the original input image. The second and third images show largest gradients mapped to a region of the image signifying the region that the model attributes to the prediction that it made, when using the Saliency Maps method. The *second row* shows localization using original class activation maps for each class represented in the model. And finally, image four in the second row is a class activation map computed with our method that localizes regions specific to our PNEUMONIA target class prediction. It can be seen that the gradient method (top row) is focusing its attention on spurious features of the image, i.e. X-ray markings (second, third, and fourth images in the top row), while our method (second row) does not have this problem. By focusing on irrelevant regions of the image (top row), the model does not provide a plausible explanation of its prediction, while by isolating a plausible region of pneumonia manifestations within the lung region, the model better explains why it made the prediction to the subject matter expert

the PNEUMONIA classification, while our method correctly localizes to the lung region, ignoring spurious activations.

7.5   Discussion

Weakly supervised localization using deep learning approaches is an active area of research. It enables machine learning practitioners and subject matter experts in various domains to increase their confidence in predictions provided by models trained on image-level labels when object-level labels are not available. We discussed several such approaches and identified possible problems when applied in real world domains. Specifically, we have shown that class overlap in training data is a source of aleatoric uncertainty and that some existing methods for weakly supervised localization tend to focus on spurious features, reducing the confidence in predictions provided by the model.

Our Amplified Directed Divergence with ensembles technique addresses the problem of class overlap by training expert binary models on classes of interest. Besides mitigating aleatoric uncertainty, our method allows the use of arbitrary, independently trained models to perform better localization, even when these models were trained on different levels of labels, i.e. image-level and object-level. The expressiveness and flexibility provide by this arbitrary pairing can provide useful feedback on the quality of model predictions in novel situations when training a new object-level label model is not feasible.

A limitation of our previous approach [34] is in the multi-class setting, specifically when more than two classes within our data can potentially overlap. Our generalization of Amplified Directed Divergence called Scaled Directed Divergence (SDD) addresses the problem for any number of overlapping classes. It mitigates the problem of neuron co-adaptation, while allowing similar data of multiple classes to be used to improve the accuracy of the overall multi-class model. In both cases, Amplified Directed Divergence and Scaled Directed Divergence improve weakly supervised localization in the presence of class overlap without spurious attention problems that can be seen in the Saliency Map approach.

Another important point to consider is how to verify obtained results. At the time of this work, and to the best of our knowledge, no spatially labeled annotations of COVID-19 regions in chest radiographs are publicly available. Once this data does become available, the standard way to validate our results of localizing relevant regions of pulmonary disease manifestations would be to use standard metrics used in supervised localization, such as intersection over union (IoU) of computed bounding boxes and ground truths bounding boxes. In addition, as more chest radiology datasets become publicly available, spatially annotated images, or object-level labels, may become available for other pulmonary diseases. This would allow us to validate our methods within the same domain, increasing the likelihood that the approach generalizes to other novel pulmonary conditions.

7.6   Conclusion

In novel situations, such as the COVID-19 pandemic, indirectly obtained image-level labels may be the only available data to aid rapid decision making at scale. We described how weakly supervised localization can improve such decision making by augmenting classifications with spatial support. We then showed that in noisy, unbalanced, and class overlapped data environments, current state-of-the-art methods for weakly supervised localization suffer from spurious attention and highly uncertain spatial localization. We described a novel technique to mitigate those challenges and empirically showed that our method improved weakly supervised localization in those environments.

Specifically, our method allows cheaper image-level labels to train classification models that can provide spatial support to standard classifications, while allowing the model to be trained on highly unbalanced and overlapped classes. The neuron co-adaptation induced by unbalanced and overlapped labels is mitigated post-training via directed divergence. The uncertainty of predictions is reduced by better isolating spatial regions responsible for target classification as shown with our natural imagery experiments. Finally, we demonstrated that our technique can be applied with minimal to no alteration to existing fundamental architectures, as shown by the use of a pre-trained ResNet152 model to improve weakly supervised localization of overlapped ImageNet classes. In addition, class activation maps computed using our method can be utilized to perform weakly supervised labeling of objects of interest in image-level labels.

As object-level labels become available in this domain, i.e. COVID-19 regions of interest labeled by radiologists, techniques to validate and improve existing methods are an important area of future work. Furthermore, as reported by Zhou et al. [43], discriminative localization can be further improved by increasing the mapping resolution of Convolutional Neural Networks used for image classification.

References

1. Ai, T., Yang, Z., Hou, H., Zhan, C., Chen, C., Lv, W., Tao, Q., Sun, Z., & Xia, L. (2020). Correlation of chest CT and RT-PCR testing in coronavirus disease 2019 (COVID-19) in China: A report of 1014 cases. *Radiology, 296*(2), 200642.
2. Alejo, R., Valdovinos, R. M., García, V., & Pacheco-Sanchez, J. H. (2013). A hybrid method to face class overlap and class imbalance on neural networks and multi-class scenarios. *Pattern Recognition Letters, 34*(4), 380–388.
3. Bernheim, A., Mei, X., Huang, M., Yang, Y., Fayad, Z. A., Zhang, N., Diao, K., Lin, B., Zhu, X., Li, K., et al. (2020). Chest CT findings in coronavirus disease-19 (COVID-19): Relationship to duration of infection. *Radiology 295*(3), 200463.
4. Bullock, J., Pham, K. H., Lam, C. S. N., Luengo-Oroz, M., et al. (2020). Mapping the landscape of artificial intelligence applications against COVID-19. arXiv preprint arXiv:2003.11336.
5. Butt, C., Gill, J., Chun, D., & Babu, B. A. (2020). Deep learning system to screen coronavirus disease 2019 pneumonia. *Applied Intelligence* 1.
6. Cao, Y., Xu, Z., Feng, J., Jin, C., Han, X., Wu, H., & Shi, H. (2020). Longitudinal assessment of covid-19 using a deep learning–based quantitative CT pipeline: Illustration of two cases. *Radiology: Cardiothoracic Imaging, 2*(2), e200082.
7. Chen, J., Wu, L., Zhang, J., Zhang, L., Gong, D., Zhao, Y., Hu, S., Wang, Y., Hu, X., Zheng, B., et al. (2020). Deep learning-based model for detecting 2019 novel coronavirus pneumonia on high-resolution computed tomography: A prospective study. MedRxiv.
8. Chowdhury, M. E., Rahman, T., Khandakar, A., Mazhar, R., Kadir, M. A., Mahbub, Z. B., Islam, K. R., Khan, M. S., Iqbal, A., Al-Emadi, N., et al. (2020). Can AI help in screening viral and COVID-19 pneumonia? arXiv preprint arXiv:2003.13145.
9. Cohen, J. P., Morrison, P., & Dao, L. (2020). COVID-19 image data collection. arXiv preprint arXiv:2003.11597.
10. Deng, J., Dong, W., Socher, R., Li, L.-J., Li, K., & Fei-Fei, L. (2009). ImageNet: A large-scale hierarchical image database. In *CVPR09*.
11. Erhan, D., Bengio, Y., Courville, A., & Vincent, P. (2009). Visualizing higher-layer features of a deep network.
12. Geirhos, R., Rubisch, P., Michaelis, C., Bethge, M., Wichmann, F. A., & Brendel, W. (2019). ImageNet-trained CNNs are biased towards texture; increasing shape bias improves accuracy and robustness. In *International Conference on Learning Representations*.

13. Ghoshal, B., & Tucker, A. (2020). Estimating uncertainty and interpretability in deep learning for coronavirus (COVID-19) detection. arXiv preprint arXiv:2003.10769.

14. Gozes, O., Frid-Adar, M., Greenspan, H., Browning, P. D., Zhang, H., Ji, W., Bernheim, A., & Siegel, E. (2020). Rapid AI development cycle for the coronavirus (COVID-19) pandemic: Initial results for automated detection & patient monitoring using deep learning CT image analysis. arXiv preprint arXiv:2003.05037.

15. Huang, L., Han, R., Ai, T., Yu, P., Kang, H., Tao, Q., & Xia, L. (2020). Serial quantitative chest CT assessment of COVID-19: Deep-learning approach. *Radiology: Cardiothoracic Imaging, 2*(2), e200075.

16. Hurt, B., Kligerman, S., & Hsiao, A. (2020). Deep learning localization of pneumonia: 2019 coronavirus (COVID-19) outbreak. *Journal of Thoracic Imaging, 35*(3), W87–W89.

17. Kanne, J. P. (2020). Chest CT findings in 2019 novel coronavirus (2019-NCOV) infections from Wuhan, China: Key points for the radiologist.

18. Krizhevsky, A., Sutskever, I., & Hinton, G. E. (2012). ImageNet classification with deep convolutional neural networks. In F. Pereira, C. J. C. Burges, L. Bottou & K. Q. Weinberger (Eds.), *Advances in neural information processing systems* (Vol. 25, pp. 1097–1105). Curran Associates.

19. Li, L., Qin, L., Xu, Z., Yin, Y., Wang, X., Kong, B., Bai, J., Lu, Y., Fang, Z., Song, Q., et al. (2020). Artificial intelligence distinguishes COVID-19 from community acquired pneumonia on chest CT. *Radiology, 296*(2), E65–E71.

20. Liang, T., et al. (2020). Handbook of COVID-19 prevention and treatment. *The First Affiliated Hospital, Zhejiang University School of Medicine. Compiled According to Clinical Experience* (Vol. 68).

21. Long, Y., Gong, Y., Xiao, Z., & Liu, Q. (2017). Accurate object localization in remote sensing images based on convolutional neural networks. *IEEE Transactions on Geoscience and Remote Sensing, 55*(5), 2486–2498.

22. Loshchilov, I., & Hutter, F. (2019). Decoupled weight decay regularization. In *International Conference on Learning Representations*.

23. Narin, A., Kaya, C., & Pamuk, Z. (2020). Automatic detection of coronavirus disease (COVID-19) using x-ray images and deep convolutional neural networks. arXiv preprint arXiv:2003.10849.

24. Oquab, M., Bottou, L., Laptev, I., & Sivic, J. (2015). Is object localization for free? Weakly-supervised learning with convolutional neural networks. In *2015 IEEE Conference on Computer Vision and Pattern Recognition (CVPR)* (pp. 685–694).

25. Ozturk, T., Talo, M., Yildirim, E. A., Baloglu, U. B., Yildirim, O., & Acharya, U. R. (2020). Automated detection of covid-19 cases using deep neural networks with X-ray images. *Computers in Biology and Medicine, 121*, 103792.

26. Qi, X., Jiang, Z., Yu, Q., Shao, C., Zhang, H., Yue, H., Ma, B., Wang, Y., Liu, C., Meng, X., et al. (2020). Machine learning-based CT radiomics model for predicting hospital stay in patients with pneumonia associated with sars-cov-2 infection: A multicenter study. medRxiv.

27. Rahman, T., Chowdhury, M., & Khandakar, A. (2020). COVID-19 radiology database. Retrieved September, 2020 from https://www.kaggle.com/tawsifurrahman/covid19-radiography-database

28. Rajpurkar, P., Irvin, J., Zhu, K., Yang, B., Mehta, H., Duan, T., Ding, D., Bagul, A., Langlotz, C., Shpanskaya, K., et al. (2017). CheXNet: Radiologist-level pneumonia detection on chest x-rays with deep learning. arXiv preprint arXiv:1711.05225.

29. Ronneberger, O., Fischer, P., & Brox, T. (2015). U-net: Convolutional networks for biomedical image segmentation. In *International Conference on Medical Image Computing and Computer-Assisted Intervention* (pp. 234–241). Springer.

30. Roy, S., Menapace, W., Oei, S., Luijten, B., Fini, E., Saltori, C., Huijben, I., Chennakeshava, N., Mento, F., Sentelli, A., et al. (2020). Deep learning for classification and localization of COVID-19 markers in point-of-care lung ultrasound. *IEEE Transactions on Medical Imaging, 39*(8), 2676–2687.

31. Shi, F., Wang, J., Shi, J., Wu, Z., Wang, Q., Tang, Z., He, K., Shi, Y., & Shen, D. (2020). Review of artificial intelligence techniques in imaging data acquisition, segmentation and diagnosis for COVID-19. *IEEE Reviews in Biomedical Engineering, 14*, 4–15.
32. Simonyan, K., Vedaldi, A., & Zisserman, A. (2014). Deep inside convolutional networks: Visualising image classification models and saliency maps. CoRR, abs/1312.6034.
33. van Sloun, R. J., & Demi, L. (2019). Localizing b-lines in lung ultrasonography by weakly supervised deep learning, in-vivo results. *IEEE Journal of Biomedical and Health Informatics, 24*(4), 957–964.
34. Verenich, E., Velasquez, A., Khan, N., & Hussain, F. (2020). Improving explainability of image classification in scenarios with class overlap: application to COVID-19 and pneumonia. In *Proceedings of the 19th IEEE International Conference on Machine Learning and Applications*.
35. Wang, L., & Wong, A. (2020). COVID-net: A tailored deep convolutional neural network design for detection of COVID-19 cases from chest x-ray images. arXiv preprint arXiv:2003.09871.
36. Wang, S., Kang, B., Ma, J., Zeng, X., Xiao, M., Guo, J., Cai, M., Yang, J., Li, Y., Meng, X., et al. (2020). A deep learning algorithm using CT images to screen for corona virus disease (COVID-19). MedRxiv.
37. Wang, X., Peng, Y., Lu, L., Lu, Z., Bagheri, M., &Summers, R. M. (2017). Chestx-ray8: Hospital-scale chest x-ray database and benchmarks on weakly-supervised classification and localization of common thorax diseases. In *2017 IEEE Conference on Computer Vision and Pattern Recognition (CVPR)* (pp. 3462–3471).
38. Won, J., Lee, S., Park, M., Kim, T. Y., Park, M. G., Choi, B. Y., Kim, D., Chang, H., Kim, V. N., & Lee, C. J. (2020). Development of a laboratory-safe and low-cost detection protocol for SARS-COV-2 of the coronavirus disease 2019 (COVID-19). *Experimental Neurobiology, 29*(2), 107.
39. Xie, X., Zhong, Z., Zhao, W., Zheng, C., Wang, F., & Liu, J. (2020). Chest CT for typical 2019-NCOV pneumonia: Relationship to negative RT-PCR testing. *Radiology, 296*(2), 200343.
40. Zhang, J., Xie, Y., Li, Y., Shen, C., & Xia, Y. (2020). COVID-19 screening on chest x-ray images using deep learning based anomaly detection. arXiv preprint arXiv:2003.12338.
41. Zheng, C., Deng, X., Fu, Q., Zhou, Q., Feng, J., Ma, H., Liu, W., & Wang, X. (2020). Deep learning-based detection for COVID-19 from chest CT using weak label. medRxiv.
42. Zheng, Z., Yao, Z., Wu, K., and Zheng, J. (2020). The diagnosis of pandemic coronavirus pneumonia: A review of radiology examination and laboratory test. *Journal of Clinical Virology, 128*, 104396.
43. Zhou, B., Khosla, A., Lapedriza, A., Oliva, A., & Torralba, A. (2016). Learning deep features for discriminative localization. In *Proceedings of the IEEE Conference on Computer Vision and Pattern Recognition* (pp. 2921–2929).
44. Zhou, Z., Siddiquee, M. M. R., Tajbakhsh, N., & Liang, J. (2018). UNet++: A nested u-net architecture for medical image segmentation. In *Deep Learning in Medical Image Analysis and Multimodal Learning for Clinical Decision Support* (pp. 3–11). Springer.

Chapter 8
A Critical Study on the Importance of Feature Selection for Diagnosing Cyber-Attacks in Water Critical Infrastructures

Ehsan Hallaji, Ranim Aljoudi, Roozbeh Razavi-Far, Majid Ahmadi, and Mehrdad Saif

8.1 Introduction

Recent advancements in cyber-physical systems and smart grids are often followed by more dependency on the application layer [10, 11, 22]. The severity of the intrusions to computer networks has been increasing continuously by threatening the security of these networks through violating privacy, integrity, and accessibility mechanisms [3, 16]. SCADA systems are used for monitoring and controlling various critical infrastructure processes [15]. While IDS monitors the attacks that occur in a system/network then processes them by detecting intrusions, it has been widely used in recent years as one of the main network security components in smart grids and cyber-physical systems [9, 22, 29]. The intrusion detection system can be characterized as a device or an application that detects malicious activities within the network.

IDS frameworks usually rely on prior knowledge, training data, or recorded data, which is often complex to analyze for extracting the attack pattern. When dealing with big data, the abundance of the recorded samples and the high dimensionality of the data, which is the focus of this work, complicate the decision-making process, as they severely decrease the efficiency of the system and quality of the constructed model. Moreover, industrial datasets usually contain noisy, redundant, or irrelevant features that introduce critical challenges to data modeling. Feature Selection (FS) techniques can be used to tackle the high dimensionality of the data and address the low quality of the data by removing redundant and non-informative features of the data [6, 13]. Such improvement in data quality will in turn enhance the performance

E. Hallaji (✉) · R. Aljoudi · R. Razavi-Far · M. Ahmadi · M. Saif
Department of Electrical and Computer Engineering, University of Windsor, Windsor, ON, Canada
e-mail: hallaji@uwindsor.ca; aljoudi@uwindsor.ca; roozbeh@uwindsor.ca; ahmadi@uwindsor.ca; msaif@uwindsor.ca

© The Author(s), under exclusive license to Springer Nature Switzerland AG 2021
M. Sayed-Mouchaweh (ed.), *Explainable AI Within the Digital Transformation and Cyber Physical Systems*, https://doi.org/10.1007/978-3-030-76409-8_8

of data-driven modules such as change detectors [24, 26] and classifiers [23, 25] in the system. FS methods have different criteria, such as their variance, entropy, and ability to preserve local similarity, which results in different correlation and consistency.

The aim of this study is to find the effect of feature selection on the detection accuracy of cyber-attacks on a cyber-physical system. To achieve this goal, we study twelve advanced feature selection models combined with two classifiers, K-Nearest Neighbors (KNN) and Decision Tree (DT). These methods are expected to effectively select the optimal set of features for detecting intrusion. The selected case study resembles the SCADA system of a water storage tank introduced in [18]. In addition, a feature analysis is performed to find the most effective features for accurate intrusion detection in the water storage system. In the context of EXplainable Artificial Intelligence (XAI), a successful feature selection process can help explaining the nature of the attack by clarifying the intrusion behind it. In other words, depending on the selected features, one can identify the system components that are more affected by the cyber-attack, which clarifies the target of the intruder.

The rest of this chapter is organized as follows. Section 8.2 describes the background of feature selection and the twelve selected feature selection techniques. Section 8.3 explains the design of an intrusion detection system that involves data-driven decision-making and analyzing data gathering. Section 8.4 presents the report on the experimental settings and results. Lastly, Sect. 8.5 will convey the conclusion of this chapter and outlines the best approach.

8.2 Background

The quality of data in a data-driven process is usually affected by various factors [5, 20, 21]. Feature selection, also known as variable selection or attribute selection, is a common technique for improving the data quality. This process obtains a subset of relevant features and eliminates the irrelevant and redundant features from the original data. The main difference between feature selection and dimensionality reduction is that the former creates a space by adopting a subset of features from the original feature space, while the latter transforms the original feature space and creates a completely new feature space. Feature selection can improve the accuracy of the model, reduce learning time, and prevent overfitting. Most feature selection methods are divided into three major buckets: (1) Filter-based: This generally analyzes intrinsic properties of data, regardless of the classifier. It only considers the association between the feature and the class label. (2) Wrapper-based: This method is based on a specific machine learning algorithm to find optimal features; it uses classifiers to score a given subset of features. (3) Embedded: This is an iterative method in which the selection process is employed into the learning of the classifier. Most of these methods can perform two processes: ranking and subset selection (sometimes they are performed sequentially); the importance of each individual feature is evaluated, usually by neglecting potential interactions among the elements of the joint set, then the final subset of features to be selected is provided.

While feature selection techniques often operate singularly and are not combined with other feature selection algorithms, it is also possible to use these techniques in combination. By doing so, one can use a simple approach such as the majority of votes to aggregate the results of these techniques. However, this approach will be most advantageous when the selected algorithms employ completely different methods (e.g., manifold learning, cluster analysis, and mutual information) to capture the distribution of the feature space. This will extend the flexibility of the feature selection process against various distribution types and data structures.

8.2.1   Infinite Feature Selection

The infinite feature selection (InfFS) is a filter-based technique that models the feature space using a graph. In this process, each graph node corresponds to a feature, and edges connecting these nodes represent pair-wise relationships between features. Weighted edges of this graphical model codify the independence between two feature distributions. A path on this graph then shows a subset of features. The convergence properties of the power series of matrices and Markov chain fundamentals are then used to evaluate the paths that contain certain features. InfFS defines a final score that shows the best feature candidate by ranking in descent order [28].

8.2.2   Infinite Latent Feature Selection

Similar to InfFS, Infinite latent feature selection (ILFS) is a probabilistic technique that models the feature space using a graph-based approach that considers all the possible subsets of features during the ranking process. However, ILFS models the relevancy between features as a latent variable in a generative process, which is inspired by the probabilistic latent semantic analysis. This enables the algorithm to investigate the feature importance upon the injection of a feature into an arbitrary set of cues [27].

8.2.3   Evolutionary Computation Feature Selection

Evolutionary computation (ECFS) has the ability to search simultaneously within a set of possible solutions to find the optimal and effective solution set, by iteratively trying to improve the feature subset with regard to a given measure of quality. An outline of three steps for EC algorithm are as follows: (1) initialization, where the population of solutions is initialized randomly; (2) evaluation of each solution in the population for fitness value; (3) iteratively generating a new population until the termination criteria (e.g., could be the maximum number of iterations or finding the optimal set of features that maximizes classification accuracy) are met [19].

8.2.4   Relief Feature Selection

Relief Feature Selection (ReliefFS) calculates a proxy statistic (referred to as feature weights) for each feature that can be used to estimate feature quality or relevance to the target concept. Relief is supplanted by ReliefFS which relies on a number of neighbors user parameter k that specifies the use of k nearest hits and k nearest misses in the scoring update for each target instance. ReliefFS finds k nearest misses from each other class and averages the weight update based on the prior probability of each class [30].

8.2.5   Mutual Information

Mutual information (MutInfFS) is a measure of dependency between two (possibly multi-dimensional) random variables that show how much knowing the value of one variable reduces the uncertainty on the others. MI is also able to capture non-linear dependencies and is invariant under invertible and differentiable transformations of the random variables in which it has been used as a score in filter methods. The selected features will be those with top mutual information w.r.t. the classes [1].

8.2.6   Maximum Relevance and Minimum Redundancy

In the Maximum Relevance and Minimum Redundancy (mRMR) method, each feature can be ranked based on its relevance to the target variable, and the ranking process is able to consider the redundancy within the selected features. The best feature is defined as one that can effectively reduce the redundant features while keeping the relevant features for the model [31].

8.2.7   Feature Selection via Concave Minimization

Feature Selection via Concave Minimization (FSV) is considered as a wrapper method in which subsets of features are sampled, evaluated, and finally kept as the final output. FSV generates a separating plane by minimizing a weighted sum of the distances of misclassified points to two parallel planes that bound the sets, and determines the separating plane midway between the set of misclassified points [2].

8.2.8   Laplacian Score

Laplacian score (Laplacian) is based on two data points that are probably related to the same topic if they are close to each other in which it is based on Laplacian Eigenmaps and Locality Preserving Projection. For each feature, the Laplacian score is computed to reflect its locality geometric structure so features that are consistent with the Gaussian Laplacian and with small weighted variance are selected [14].

8.2.9   Multi-Cluster Feature Selection

Multi-Cluster Feature Selection (MCFS) uses a multi-cluster structure that is defined to measure the correlations between different features without label information (unsupervised feature selection). Recently, the spectral clustering structure of the data shows a significant interest in which data points are structured using the top eigenvectors of graph Laplacian (manifold learning) and find the subset selection using L1-regularized models [4].

8.2.10   Recursive Feature Elimination

Recursive Feature Elimination (RFE) is basically a recursive process that ranks features according to some measure of their importance. The less relevant feature is removed iteratively since it has the least effect on the classification. Therefore, RFE aims to eliminate dependencies and collinearity that may exist in the model. For high correlated features and large data sets, the relative importance of each feature can change substantially when analyzed over a different subset of features during the stepwise elimination process in which recursion is used [7].

8.2.11   L0-Norm

Norms are a way to measure size or length in higher dimensions. L0-norm is the most direct and ideal scheme of feature selection that is difficult to optimize so L0-norm can balance the training error against the number of non-zero features [12].

L0-Norm penalizes features by which the regularization and parallel parameter estimation processes become more complicated. L0-norm solves the L0 penalty problem by selecting non-zero coefficients and regularization parameters simultaneously. In addition, it finds an estimated solution for this penalty problem.

8.2.12 Fisher Score

Fisher score finds a subset of feature, which selects the top-ranked features with large scores. The score of each feature is computed independently by the heuristic algorithm. The algorithm fails to select features that have low individual scores but a very high score when they are combined together as a whole [8].

8.3 Design of Intrusion Detection System

The designed Intrusion detection system (IDS) uses a multi-modular structure, in which the traffic data initially passes the FS methods. Then, the selected features of data will be passed to the classification module, where the normal and malicious traffic can be classified based on their type (see Fig. 8.1).

8.3.1 Data Collection

The data is collected from a cyber-physical system that resembles a water storage tank [17]. SCADA systems collect data from remote facilities about the state of

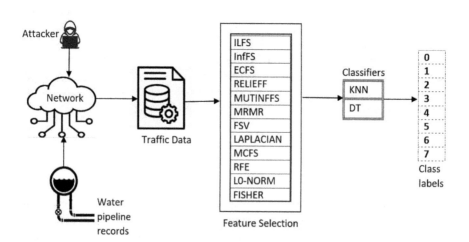

Fig. 8.1 Illustrative diagram of the designed intrusion detection systems

the physical process and send commands to control the physical process creating a feedback control loop. SCADA was used to control a water storage tank as it has communication patterns that are set of repetitive read and write commands. First, it writes the contents of all registers and coils used for control. Then, a MODBUS protocol reads the holding register command that measures the state of the system. This protocol acts as a single serial cable that connects the serial ports on Master and Slave devices. These two commands are each followed by a response. The raw collected data consisted of variables such as command and response address, command and response memory, command and response memory count, command read and write function, the response read and write function, subfunction, command length, response length, control mode, control scheme, high (H) set-point value, high alert (HH) set-point value, low (L) set-point value, low alert (LL) set-point value, pump state, cyclic redundancy code (CRC) error rate, water level measurement, timestamps, and attack class.

8.3.2 Decision-Making

In our case study, the DT and KNN algorithms are developed as classifiers in combination with feature selection techniques. A decision tree is a tree-like graph consisting of internal nodes that represent a test on an attribute and branches, which denote the outcome of the test and each leaf node holds a class label. K-nearest neighbors is a supervised metric learning algorithm that use the label information to learn a new unlabeled data based on a similarity measure by calculating the distance between points using distance measures such as Euclidean distance, Hamming distance, Manhattan distance, and Minkowski distance. Decision trees and KNN can analyze data and identify significant characteristics in the network that indicate malicious attacks.

The intrusion detection system detects and classifies seven different types of cyber-attacks in the water storage system, including a normal class when the system is safe, as shown in Table 8.1. These injection attacks are also explained briefly in the following:

- Naïve Malicious Response: can be used to send fake payloads by injecting response packets into the network.
- Complex Malicious Response: conceals the state of the controlled physical process to maliciously affect the feedback control loop.
- Malicious State Command: manipulates remote field devices to change the normal system state to a critical state by sending malicious commands.
- Malicious Parameter Command: mainly tries to change the set-points defined for programmable logic controllers.
- Malicious Function Code: refers to the commands included in the application layer of a system, which can be used maliciously by attackers to create unintended consequences.

Table 8.1 List of simulated cyber-attacks in water pipeline system

Classes	Types of attacks
0	Instance not part of an injection
1	Naïve malicious response injection
2	Complex malicious response injection
3	Malicious state command injection
4	Malicious parameter command injection
5	Malicious function code injection
6	Denial of service injection
7	Reconnaissance injection

- Denial of Service: corrupts communications links and system programs by attempting to exhaust computational resources.
- Reconnaissance: is the process in which attackers gain device information and system vulnerabilities to plan future attacks against a SCADA system.

The network traffic data are validated and trained by incorporating state-of-the-art feature selection techniques into an intrusion detection system that consists of different modules. After the traffic data is passed through feature selection techniques, the most relevant features are selected and new reduced data is trained using KNN and DT classifiers. The reduced data is classified into seven different classes as well as the safe state class, as shown in Fig. 8.1.

8.4 Experimental Results

This section analyses the obtained results in terms of accuracy and standard deviation and compares it with twelve different FS techniques. Figure 8.1 shows all the feature selection techniques that were used in our approach.

8.4.1 Experimental Setting

The water storage tank system generates network flow records that are captured with a serial port data logger that includes 200,000 samples recorded using a laboratory-scale test bed. An imbalanced data set was randomly selected by sampling 10% (27199 samples) of the instances to minimize memory requirements and the processing time. 19503 of these samples correspond to the normal state (i.e., class 0), and the rest of the samples are collected when the system was under attack. Classes 1–7 in Table 8.1 have 1198, 1457, 209, 410, 155, 135, and 4132 samples, respectively. In order to detect malicious activities in the water storage system, features were divided into network traffic features and payload content features. The former gives information regarding the communications within the SCADA

Table 8.2 List of raw features in the water storage system

Number	Feature Name	Description
1	Command address	Device ID in command packet
2	Response address	Device ID in response packet
3	Command memory	Memory start position in command packet
4	Response memory	Memory start position in response packet
5	Command memory count	Number of memory bytes for R/W command
6	Response memory count	Number of memory bytes for R/W response
7	Command read function	Value of read command function code
8	Command write function	Value of write command function code
9	Response read function	Value of read response function code
10	Response write function	Value of write response function code
11	Sub-function	Value of sub-function code in the command/response
12	Command length	Total length of command packet
13	Response length	Total length of response packet
14	H	Value of H set-point
15	HH	Value of HH set-point
16	L	Value of L set-point
17	LL	Value of LL set-point
18	Control mode	Automatic, manual, or shutdown
19	Control scheme	Control scheme of the water pipeline
20	Pump	Value of pump state
21	CRC rate	CRC error rate
22	Measurement	Water level
23	Time	Time interval between two packets
24	Result	Manual classification of the instance

network system, while the latter describes the current state for different components of the SCADA system. The developed data set consists of 24 unique features (i.e., 8 payload and 16 network traffic features) as shown in Table 8.2.

The data set described in this chapter used MODBUS traffic from RS-232 connection in which it is one byte long with each server having a unique device address. The water storage tank holds approximately two liters of water that consists of a relief valve to drain water from the tank, a pump to add water to the tank, and a meter to measure the percentage of water level. In addition to the on/off control scheme to maintain the water level between high (H) and low (L) set-points, an alarm is turned on when the water level is above high alarm set-point (HH) or below the low alarm set-point (LL).

The read and write commands/responses have a fixed length for each system in which an attack can be observed. In a normal system, the error rate of the intrusion detection (i.e., w.r.t. accuracy and F1-score) should be low and constant, but when the system is subjected to a denial-of-service attack the rates are anticipated to increase. During the normal state, if there is no error, the response function code matches the command function code. In the presence of an error, the response sub-function code is changed to the command function code plus a value of 0X80. When a sensor detects the water level has reached L (H) level, the programmable logic control turns the water pump on (off).

In order to log the data and inject attacks, a bump-in-the-wire method is used. The device implementation is conducted using C programming and VMware virtual machine. Two RS-232 serial ports are included in the virtual machine that are connected to a USB-to-serial converter. The programmed software monitors serial ports for traffic. Any detected traffic is then timestamped and saved in a log file. Furthermore, the software incorporated hooks to inject, delay, drop, and alter network traffic to facilitate the attacks.

8.4.2 Results Analysis

Results were analyzed based on 10-fold cross-validation iterations for each feature selection technique, as shown in Fig. 8.2, to divide the outcomes w.r.t. KNN and DT classifiers.

Figure 8.2 shows the performance measure in terms of accuracy through cross-validation and by resorting to feature selection techniques along with the KNN and DT classifiers. Before the feature selection is being operated on the dataset, DT classifier displays better results than KNN classifier. However, after the feature selection, KNN classifier shows a satisfying improvement and recorded a higher average accuracy than DT classifier.

In Figs. 8.3 and 8.4, it is conspicuous that ECFS, INfFS, and ILFS methods show higher performance in terms of accuracy and F1-score. The accuracy performance on KNN classifier is demonstrated in Fig. 8.3a in which the ECFS method illustrates the best results among all FS methods, and its average accuracy is approximately

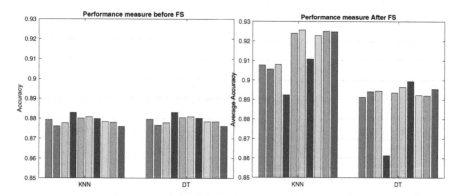

Fig. 8.2 Classification accuracy of DT and KNN classifiers before and after the FS. Each bar shows the obtained accuracy for one iteration of the 10-fold cross-validation.

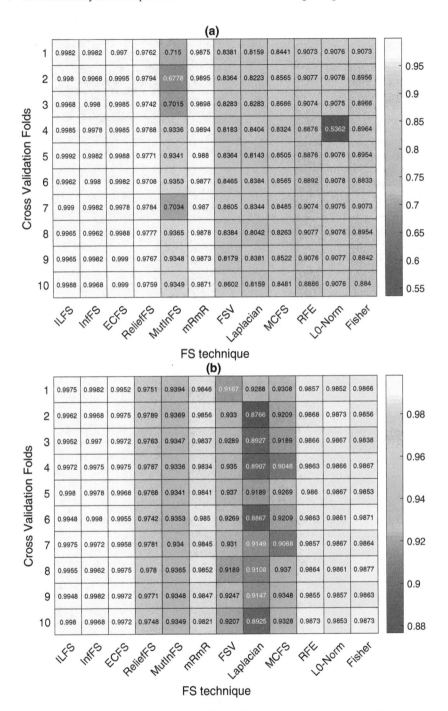

Fig. 8.3 Accuracy profile of the FS methods in different iterations of cross-validation, w.r.t. KNN and DT classification results. (**a**) Accuracy performance on KNN classifier. (**b**) Accuracy performance on DT classifier

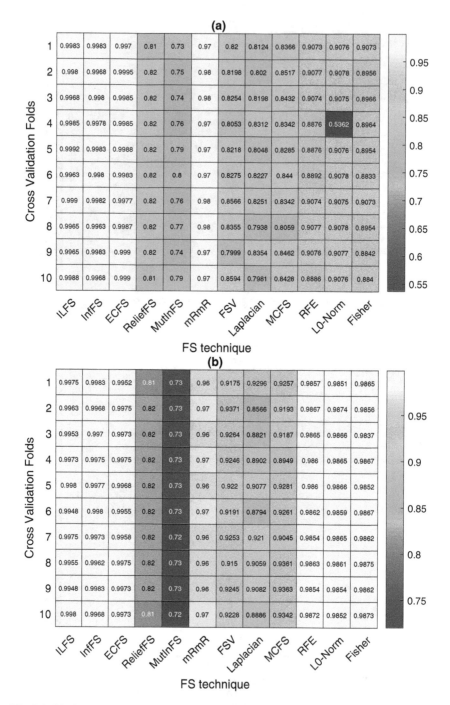

Fig. 8.4 Obtained F1-Scores by FS methods in different iterations of cross-validation, w.r.t. KNN and DT classification results. (**a**) F1-score performance on KNN classifier. (**b**) F1-score performance on DT classifier

99.85%. Furthermore, ILFS and InfFS are ranked second and third with an average of 99.7%, and it is likewise considering the F1-score performance on KNN, as shown in Fig. 8.4b. Moreover, mRMR, ReliefF, RFE, Fisher, L0-Norm, MCFS, MutInFS, FSV, and Laplacian methods are ranked from fourth to 12-th, while their accuracy performance falls between 82% and 98% on the KNN classifier.

The presentation of accuracy performance using the DT classifier is shown in Fig. 8.3b in which InfFS method has outperformed the other FS methods with an accuracy measure of 99.7%. ECFS and ILFS methods come in the second and third ranks with an average of 99.6%. The rest of the methods are ranked from fourth to 12-th and sorted as: Fisher, RFE, L0-norm, mRMR, ReliefFS, MutInfFS, FSV, MCFS, and Laplacian. ECFS, ILFS, and InfFS methods result in the best accuracy compared with the other nine FS methods. In addition, the Laplacian method is less likely to be sensitive to the choice of classifiers.

To study the F1-score on KNN and DT classifiers, Fig. 8.4 illustrates the results of the twelve utilized FS techniques. Considering the results of the KNN classifier, ECFS recorded the highest F1-score when combined with the KNN and DT classifiers in Fig. 8.4a, b, respectively. ILFS and InfFS maintain their second and third ranks when coupled with KNN classifier. Using DT, however, changes InfFS and ILFS ranks to second and third, respectively. Moreover, in respect of KNN classifier, mRMR, RFE, Fisher, L0-Norm, MCFS, FSV, ReliefF, Laplacian, and MutInfFS methods are ranked from fourth to 12-th, respectively. Considering the DT classifier, FS methods are ranked in the following order from fourth to the 12-th rank: Fisher, RFE, L0-norm, mRMR, FSV, MCFS, laplacian, REliefF, MutInfFS. ECFS, ILFS, and InfFS methods are more stable and always improve the F1-score. MutInFS method has failed to improve accuracy and F1-score performance.

Figure 8.5 shows the F1-score performance on ECFS, ILFS, and InfFS methods, which are more compatible with the KNN classifier. This is while the rest of FS methods are suggested to be used along with DT classifier. In general, ECFS performs better than other FS techniques and results in the maximum accuracy when coupled with KNN that is about 99.85%. In addition, MutInFS worsens the F1-score; however, stability is improved when it is used with KNN, which is about 76.3%. RFE, L0-norm, and Fisher techniques result in a stable and a slight difference in F1-score when coupled with DT classifier in which it scores close to 98%.

8.4.3 Feature Analysis

Considering the FS outputs for all the studied FS methods, Fig. 8.6 illustrates the importance of each feature w.r.t. the number of times it is selected by FS methods. Based on the results shown in this figure, it can be inferred that the most important features are the response address and time (number 2 and 23 in Table 8.2). The second most important features are features 9 and 15, which are response read function and HH, respectively. The third group of important features

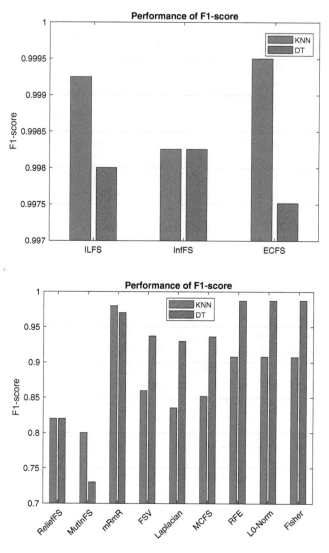

Fig. 8.5 Averaged F1-score of FS methods over the 10-fold cross-validation using DT and KNN

are command address, command memory, and command length (numbers 1, 3, and 12 in Table 8.2). These seven features are selected more than six percent of the times and they are believed to be most effective on the detection accuracy. This while the least informative features for intrusion detection of the water storage tank seem to be control scheme and pump, which are selected only one percent of the times.

The information obtained from the feature selection algorithms can be used to explain the nature of the attack, which in turn helps to plan a suitable response or counter-attack. For instance, one of the top features in Fig. 8.6, namely response

Feature Importance

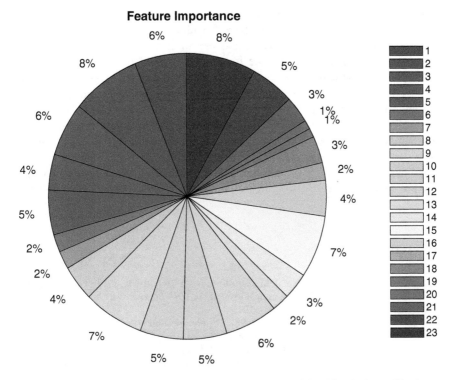

Fig. 8.6 Importance of features based on the overall results of the FS techniques. The feature numbers correspond to the list of features in Table 8.2

address (numbers 2) can be used to detect the reconnaissance attack, as the mismatch between response device addresses is usually an indicator of this attack. Another example is another top feature, time (number 23 in Fig. 8.6), which can be used to detect three types of cyber-attacks such as malicious command injection, malicious response injection, and DOS attacks. The time interval between packets is almost consistent during the normal operation; however, this measurement becomes very different when such attacks exist in the network. Therefore, one can explain the nature of attack of detecting anomalies in any of these features, as they are indicator of certain events. Knowing the most important features, on the other hand, can inform us regarding the most targeted parts of the system, and its mechanism, which is useful for planning and taking defensive actions.

8.5   Conclusion

In this chapter, twelve feature selection techniques are reviewed and analyzed on a cyber-physical case study. The selected case study resembles a SCADA system implemented for a water storage tank, which is under cyber-attacks. The selected feature selection techniques are employed within a multi-modular IDS, which combines a set of feature selection techniques with two classifiers. This framework enables a comparative study on the feature selection methods, as well as their compatibility with the selected classifiers. Moreover, a feature analysis is performed w.r.t. the results of the feature selection that determines the most important features that are crucial for the task of intrusion detection in the given SCADA system. The features selection methods in this study achieved satisfying results in terms of accuracy and F1-score. The results indicate that feature selection could improve some certain level of classification accuracy in IDS. The performed comparative experiment suggests the best combination of feature selection algorithm with a classifier and suggests which features should be included in the detection model.

References

1. Beraha, M., Metelli, A. M., Papini, M., Tirinzoni, A., & Restelli, M. (2019). Feature selection via mutual information: New theoretical insights. In *2019 International Joint Conference on Neural Networks (IJCNN)* (pp. 1–9). IEEE.
2. Bradley, P. S., & Mangasarian, O. L. (1998). Feature selection via concave minimization and support vector machines. In *ICML* (Vol. 98, pp. 82–90).
3. Buczak, A. L., & Guven, E. (2016). A survey of data mining and machine learning methods for cyber security intrusion detection. *IEEE Communications Surveys Tutorials, 18*(2), 1153–1176.
4. Cai, D., Zhang, C., & He, X. (2010). Unsupervised feature selection for multi-cluster data. In *Proceedings of the 16th ACM SIGKDD International Conference on Knowledge Discovery and Data Mining* (pp. 333–342).
5. Farajzadeh-Zanjani, M., Razavi-Far, R., & Saif, M. (2016). Efficient sampling techniques for ensemble learning and diagnosing bearing defects under class imbalanced condition. In *IEEE Symposium Series on Computational Intelligence (SSCI)* (pp. 1–7).
6. Farajzadeh-Zanjani, M., Razavi-Far, R., & Saif, M. (2018). A critical study on the importance of feature extraction and selection for diagnosing bearing defects. In *IEEE 61st International Midwest Symposium on Circuits and Systems (MWSCAS)* (pp. 803–808).
7. Granitto, P. M., Furlanello, C., Biasioli, F., & Gasperi, F. (2006). Recursive feature elimination with random forest for PTR-MS analysis of agroindustrial products. *Chemometrics and Intelligent Laboratory Systems, 83*(2), 83–90.
8. Gu, Q., Li, Z., & Han, J. (2012). Generalized fisher score for feature selection. arXiv preprint arXiv:1202.3725.
9. Hallaji, E., Razavi-Far, R., & Saif, M. (2020). Detection of malicious SCADA communications via multi-subspace feature selection. In *International Joint Conference on Neural Networks (IJCNN)* (pp. 1–8).
10. Hammami, Z., Sayed Mouchaweh, M., Mouelhi, W., & Ben Said, L. (2018). Discussion and review of the use of neural networks to improve the flexibility of smart grids in presence of distributed renewable resources. In *2018 17th IEEE International Conference on Machine Learning and Applications (ICMLA)* (pp. 1304–1309).
11. Hammami, Z., Sayed Mouchaweh, M., Mouelhi, W., & Ben Said, L. (2020). Neural networks for online learning of non-stationary data streams: a review and application for smart grids flexibility improvement. *Artif Intelligence Review, 53*, 6111–6154.

12. Han, J., Sun, Z., & Hao, H. (2015). *l0*-norm based structural sparse least square regression for feature selection. *Pattern Recognition, 48*(12), 3927–3940.
13. Hassani, H., Hallaji, E., Razavi-Far, R., Saif, M.: Unsupervised concrete feature selection based on mutual information for diagnosing faults and cyber-attacks in power systems. *Engineering Applications of Artificial Intelligence, 100*, 104150 (2021)
14. He, X., Cai, D., & Niyogi, P. (2006). Laplacian score for feature selection. In *Advances in neural information processing systems* (pp. 507–514).
15. Igure, V. M., Laughter, S. A., & Williams, R. D. (2006). Security issues in SCADA networks. *Computers & Security, 25*(7), 498–506.
16. Liao, H. J., Richard Lin, C. H., Lin, Y. C., & Tung, K. Y. (2013). Intrusion detection system: A comprehensive review. *Journal of Network and Computer Applications, 36*(1), 16–24.
17. Morris, T., & Gao, W. (2014). Industrial control system traffic data sets for intrusion detection research. In *International Conference on Critical Infrastructure Protection* (pp. 65–78). Springer.
18. Morris, T., Srivastava, A., Reaves, B., Gao, W., Pavurapu, K., & Reddi, R. (2011). A control system testbed to validate critical infrastructure protection concepts. *International Journal of Critical Infrastructure Protection, 4*(2), 88–103.
19. Nakisa, B., Rastgoo, M. N., Tjondronegoro, D., & Chandran, V. (2018). Evolutionary computation algorithms for feature selection of EEG-based emotion recognition using mobile sensors. *Expert Systems with Applications, 93*, 143–155.
20. Razavi-Far, R., Cheng, B., Saif, M., & Ahmadi, M. (2020). Similarity-learning information-fusion schemes for missing data imputation. *Knowledge-Based Systems, 187*, 104805.
21. Razavi-Far, R., Farajzadeh-Zanjani, M., & Saif, M. (2017). An integrated class-imbalanced learning scheme for diagnosing bearing defects in induction motors. *IEEE Transactions on Industrial Informatics, 13*(6), 2758–2769.
22. Razavi-Far, R., Farajzadeh-Zanjani, M., Saif, M., & Chakrabarti, S. (2020). Correlation clustering imputation for diagnosing attacks and faults with missing power grid data. *IEEE Transactions on Smart Grid, 11*(2), 1453–1464.
23. Razavi-Far, R., Hallaji, E., Farajzadeh-Zanjani, M., & Saif, M.: A semi-supervised diagnostic framework based on the surface estimation of faulty distributions. *IEEE Transactions on Industrial Informatics, 15*(3), 1277–1286.
24. Razavi-Far, R., Hallaji, E., Saif, M., & Ditzler, G. (2019). A novelty detector and extreme verification latency model for nonstationary environments. *IEEE Transactions on Industrial Electronics, 66*(1), 561–570.
25. Razavi-Far, R., Hallaji, E., Saif, M., & Rueda, L. (2017). A hybrid scheme for fault diagnosis with partially labeled sets of observations. In *16th IEEE International Conference on Machine Learning and Applications (ICMLA)* (pp. 61–67).
26. Razavi-Far, R., & Kinnaert, M. (2012). Incremental design of a decision system for residual evaluation: a wind turbine application. *IFAC Proceedings Volumes, 45*(20), 343–348. 8th IFAC Symposium on Fault Detection, Supervision and Safety of Technical Processes
27. Roffo, G., Melzi, S., Castellani, U., & Vinciarelli, A. (2017). Infinite latent feature selection: A probabilistic latent graph-based ranking approach. In *Proceedings of the IEEE International Conference on Computer Vision* (pp. 1398–1406).
28. Roffo, G., Melzi, S., & Cristani, M. (2015). Infinite feature selection. In *Proceedings of the IEEE International Conference on Computer Vision* (pp. 4202–4210).
29. Sayed-Mouchaweh, M. (Ed.) (2018). *Diagnosability, security and safety of hybrid dynamic and cyber-physical systems* (1st ed). Springer.
30. Urbanowicz, R. J., Meeker, M., La Cava, W., Olson, R. S., & Moore, J. H. (2018). Relief-based feature selection: Introduction and review. *Journal of biomedical informatics, 85*, 189–203.
31. Zhao, Z., Anand, R., & Wang, M. (2019). Maximum relevance and minimum redundancy feature selection methods for a marketing machine learning platform. arXiv preprint arXiv:1908.05376.

Chapter 9
A Study on the Effect of Dimensionality Reduction on Cyber-Attack Identification in Water Storage Tank SCADA Systems

Ranim Aljoudi, Ehsan Hallaji, Roozbeh Razavi-Far, Majid Ahmadi, and Mehrdad Saif

9.1 Introduction

Abundance of features in cyber-physical systems can complicate explaining various events for various applications. Security challenges arise from two different perspectives. Firstly, an event may be detectable, when a change happens in a certain combination of features, and as the number of these variables and events increases, explaining the system status becomes more difficult. Furthermore, the feature space may contain hidden characteristics that are dormant to human eye. Devising dimensionality reduction technique improves the explainability of such systems in a number of ways. These techniques aim at improving the feature space by capturing the complex structure of the original data and then transform it into a low-dimensional space, which facilitates visualization, revealing relationships between samples, understanding and monitoring the dynamics of the system.

Dimensionality reduction can be very helpful in the design of intrusion detection systems (IDS). For instance, if a cyber-attack can be detected by monitoring a large number of features, dimensionality reduction can yield a feature space in which only one or a very small number of features are enough to explain a change that indicates a cyber-attack [7]. In contrast, other techniques such as feature selection may not result in the same efficiency, as the features may not have enough information to only select a small number of them to detect a cyber-threat. In other words, feature selection usually works when at least a number of features possess very useful information, while dimensionality reduction tries to rectify the feature space and obtain an improved distribution.

R. Aljoudi · E. Hallaji (✉) · R. Razavi-Far · M. Ahmadi · M. Saif
Department of Electrical and Computer Engineering, University of Windsor, Windsor, ON, Canada
e-mail: aljoudi@uwindsor.ca; hallaji@uwindsor.ca; roozbeh@uwindsor.ca; ahmadi@uwindsor.ca; msaif@uwindsor.ca

An IDS is a security mechanism to inspect traffic via detecting and tackling computer security threats or any suspicious behavior [4, 16]. Challenges are arising in accurately detecting intrusions, which make the majority of studies on smart grids and cyber-physical systems to focus on more advanced approaches such as machine learning [24, 31]. To secure industrial network systems, such as smart grids [9, 10, 24], we require to address malicious intrusions that are violating privacy, integrity, and accessibility.

SCADA systems are used for monitoring and controlling various critical infrastructure processes through receiving data from sensors [8, 13]. It controls the mechanical machines, while the software allows human interactions to manage the machines. A traditional IDS needs a database that holds records of different attacks, each record corresponds to a particular intrusion and its characteristics. The major drawback of this mechanism is the necessity for human involvement to inspect threats, which is a very complicated and time consuming task. Thus, machine learning techniques promote anomaly detection algorithms that can discover abnormal changes in the system [8, 28, 29]. For example, in our study, we are elaborating dimensionality reduction techniques in an intrusion detection system to optimize the performance of security mechanisms.

A challenging task for data modeling is to apply IDS on high-dimensional data streams as a high volume of data may include noisy, redundant, or irrelevant features [27]. Therefore, irrelevant and redundant features reduce the quality of the learning process and grow the risk of the classifier to over-fit. To eliminate this issue, a subset of relevant features is often selected to construct a strong learning model. In addition, the high-dimensionality of the dataset affects the prediction accuracy of machine learning algorithms and data visualization. Reducing the dimensionality improves the classification performance [26], which in turn enhances the robustness of intrusion detection systems.

In this study, we will explore the effect of dimension reduction (DR) on detection accuracy of cyber-attacks on cyber-physical system. We experiment 22 advanced feature extraction models combined with two classifiers, K-Nearest Neighbors (KNN) and Decision Tree (DT), which is expected to effectively select the optimal set of features in detecting intrusion. A water storage tank system introduced by Morris et al. [17] is applied.

The remainder of this study is structured as follows. Section 9.2 briefly introduce the background of dimension reduction and dimension reduction methods. The Sect. 9.3 explains the design of the intrusion detection system that involves data-driven decision making and analyzing data gathering. Section 9.4 reports the experimental settings and results. Lastly, Sect. 9.5 outlines the conclusion of this chapter.

9.2   Background

Industrial data-driven models are often challenged with various obstacles [22, 23, 25]. One of the most common challenges in machine learning is the issue of high-dimensionality, which can be addressed by dimensionality reduction. Dimensionality reduction is the process of improving the original feature space and transforming it into a smaller one in order to minimize the complexity of a model and avoid curse of the dimensionality [5, 7]. Dimensionality reduction is mostly used for data analysis, compression, and data visualization. Most of the feature reduction methods are divided into two main categories: (i) Feature selection: approaches select a subset of features from the original feature space that results in the optimal performance. (ii) In contrast, dimensionality reduction, also called feature extraction, captures the structure of the original feature space and then transforms into a lower-dimensional features space. This data transformation may be linear or non-linear. The focus of this work is on the dimensionality reduction, and the selected techniques are explained in the following subsections.

In contrast to feature selection techniques that may be used in combination to provide different rankings for the features, dimensionality reduction techniques are preferred to be used alone. This is due the fact that the created features spaces may represent different distributions and do not share any common features. A question, however, may arise regarding the criteria for selecting the right technique for the task at hand. While various measures can be used to facilitate this decision, the best choice could be made after testing different algorithms on the same data to see which one is more adaptable with the case study and bring about higher performance. This is the approach followed in this chapter. Nevertheless, should one desire to choose a versatile technique that works with various case studies, there are a few points to consider. Firstly, it is more desirable to use supervised dimensionality reduction methods, if labeled data is available, as their valuable information will be discarded by unsupervised methods. Secondly, many of dimensionality reduction methods make use of approaches such as manifold learning and kernel functions. These techniques are very powerful, if they are carefully optimized, and the data distribution should match the underlying assumptions such as comparability with the selected kernels or to be projectable onto a manifold.

9.2.1   Principal Component Analysis

Principle Component Analysis (PCA) is a very established method, as an unsupervised linear transformation technique. PCA supports us to identify patterns in the data based on the correlation between features. PCA projects the direction of maximum variance in high-dimensional data onto a lower-dimensional subspace in order to minimize the sum of squared error, or maximize the variance. It is decomposed by obtaining eigenvectors and eigenvalues on the data covariance matrix of the whole dataset. The obtained eigenvalues represent the variance of the projected inputs along principal axes, and eigenvectors (principal components) determine the directions of the new feature space. The benefits of PCA include the reduction of noise in the data and the ability to produce independent and uncorrelated features [33].

9.2.2 Factor Analysis

Factor Analysis (FA) is a statistical method that can be considered as an extension of PCA. FA is designed to identify the unobservable variables from the observed patterns of correlation between the variables. This is in contrast to the PCA, as it is unable to use the observed information. A factor is correlated with multiple observed variables, so each factor describes an appropriate amount of variance in the observed variables [18].

9.2.3 Confirmatory Factor Analysis

Confirmatory Factor Analysis (CFA) is a multivariate statistical method that measures variables representing the number of constructs (or factors). CFA models the data density on a low-dimensional manifold on which the data is representable [32]. CFA also follows a global approach for parameter optimization of the manifold estimation, which results in a satisfying convergence rate.

9.2.4 Multidimensional Scaling

Multidimensional Scaling (MDS) references the overall similarity (or dissimilarity) of the objects. MDS is used to visualize the dissimilarities or distances (usually by Euclidean distance) between objects by projecting the points to a low-dimension space [14].

9.2.5 Linear Discriminant Analysis

Linear Discriminant Analysis (LDA) is a supervised linear transformation that reduces the dimensionality on multi-class data by linearly projecting the original samples to a smaller space, while maintaining the class-discriminatory characteristics of the original data [21].

9.2.6   Isomap

Isomap (ISO) is also referred to as isometric mapping; it is a non-linear dimensionality reduction method, which takes advantage of local information by using the concept of geodesic distances induced by a neighborhood graph. This graph is embedded between pairs of points rather than Euclidean distances [30].

9.2.7   Semantic Mapping

Semantic Mapping (SM) reduces the dimensionality by clustering the original co-occurrent features. Using these semantic clusters and combining features mapped in the same cluster, it then generates an extracted feature that contains semantically related terms [1].

9.2.8   Probabilistic Principal Component Analysis

Probabilistic Principal Component Analysis (PPCA) offers an extension to the scope of PCA. PPCA can be utilized as a Gaussian model by maximizing the likelihood estimates of the parameters that are associated with the covariance matrix and can be efficiently computed from the data principle component [35].

9.2.9   Locally Linear Embedding

Locally Linear Embedding (LLE) is an unsupervised learning algorithm and a non-linear dimensionality reduction technique. LLE outlines its inputs into a single global coordinate system of lower-dimensionality without the involvement of local minima. By employing the local symmetries of linear reconstructions, it can study the global structure of non-linear manifolds. LLE projects the points to a locally linear neighborhood. LLE utilizes an eigenvector based optimization technique to find the low-dimensional embedding of points [33].

9.2.10   Laplacian Eigenmaps

Laplacian Eigenmaps (LE) is a closely related approach to LLE. LE constructs a graph to compute a low-dimensional representation of the dataset that preserves local neighborhood constraints of the dataset in an optimal process. LE is constructed by a weighted graph with k nodes. Each data point is a node, and a set of edges connecting the proximity of neighboring points using the K-nearest neighbor algorithm [3].

9.2.11 Landmark Isomap

Landmark Isomap (LIM) is a variant of Isomap that selects a group of points termed as landmarks to simplify the embedding computation. LIM only computes the shortest path from each data point to the landmark points. The classical MDS is then applied to the resulting geodesic distance matrix to find a Euclidean low-dimension embedding of all data points [19].

9.2.12 Hessian-based Locally Linear Embedding

Hessian-based Locally Linear Embedding (HLLE) may be considered as an improved version of the LLE. Its theoretical approach is somehow similar to the Laplacian eigenmap framework, if the Laplacian operator is replaced with the Hessian. HLLE uses orthogonal coordinates on the tangent planes. This makes the local fits more robust for the dimensionality reduction [6].

9.2.13 Local Tangent Space Alignment

Local Tangent Space Alignment (LTSA) uses manifold learning, which can convert a non-linear embedding of high-dimensional data into a smaller space and rebuild high-dimensional coordinates from embedding coordinates. The steps for performing LTSA are similar to LLE; however, it is different in optimizing the embedding. In LTSA, we compute the tangent space of each data point and align those local tangent spaces, while ignoring the label information [37].

9.2.14 Kernel Principal Component Analysis

Kernel PCA (KPCA) is an extension of PCA for performing non-linear dimensionality reduction through the use of kernels. PCA can be applied to datasets that are linearly separable. This is while kernel PCA maps non-linear datasets and uses a kernel function (also called non-linear mapping function) to project dataset onto a higher dimensional feature space, where it is linearly separable [15].

9.2.15   Generalized Discriminant Analysis

Generalized Discriminant Analysis (GDA) is designed for a non-linear transformation. It utilizes kernel functions to map the data onto a new space, which leads to non-linear discriminant analysis for the input data. This has been done by maximizing the ratio of the between-class scatter to the within-class scatter [2].

9.2.16   Neighborhood Preserving Embedding

Neighborhood Preserving Embedding (NPE) is a linear DR method that aims to discover the local neighborhood structure on the data manifold. Each data point is represented as a linear combination of the neighboring data points and coefficients that are specified in the weight matrix. It then finds an optimal embedding such that the neighborhood structure can be preserved in the resulted feature space [11].

9.2.17   Locality Preserving Projections

Locality Preserving Projections (LPP) is similar to NPE in aiming at preserving the local manifold structure. LPP shares a lot of LE or LLE properties. LPP employs the concept of non-linear Laplacian eigenmap and computes a transformation matrix that maps the data points to a new space. The projective maps in LPP are the optimal linear approximations to the eigenfunctions of the Laplace Beltrami operator on the manifold [12].

9.2.18   Diffusion Maps

Diffusion Maps (DM) reduces the data dimensionality by re-arranging data according to parameters of its underlying manifold. The Euclidean distance between points in the embedded space is equal to the diffusion distance in the original dimension space. The connectivity between the points is measured using a local similarity measure at different scales [20].

9.2.19   Locally Linear Coordination

Locally Linear Coordination (LLC) computes a number of locally linear models on data using the Expectation Maximization approach. By this mean, it performs a global alignment of the linear models by aligning the local linear models using a variant of LLE [36].

9.2.20 Manifold Charting

Manifold Charting (MC) minimizes a cost function that measures the amount of difference between the linear models on the global coordinates of the data points by solving a generalized eigenproblem [36]. MC also shares some similarities with the LLC technique.

9.2.21 Large Margin Nearest Neighbor

Large Margin Nearest Neighbor (LMNN) is based on semi-definite programming for optimizing a convex problem. The target neighbors can be set as a k-nearest neighbors rule that shares the same labeled instances. The new data instances are obtained from the highest vote of the k closest labeled instances. Using the global distance metric learning method, it measures the nearby instances from the same class and eliminates instances from different classes [34].

9.2.22 Independent Component Analysis

Independent Component Analysis (ICA) is a computational method that transforms the independent components of the observed data by increasing the statistical independence of the estimated components. ICA aims to separate multivariate signals into components that are maximally independent of each other by applying a linear transformation to decompose the original data. ICA aims to increase the accuracy for uncorrelated data; however, the obtained independent components may be irrelevant [33].

9.3 Design of Intrusion Detection System

The aim of the intrusion detection system (IDS) is to monitor and secure the industrial control system from malicious network traffic. This ensures the integrity and confidentiality of streaming data and the availability of services as well.

9.3.1 Data Collection

The SCADA datasets have been used for the evaluation of the intrusion detection. In this case study, SCADA was implemented for a water storage tank system. An intruder can hack into the network system of this cyber-physical system and disrupt the operation of the control unit. SCADA systems are generally made of four group of components. The first components are the sensors and actuators that collect data from remote facilities. These data have information about the state of the physical process. By this mean, commands can be sent to control the physical process and create a feedback control loop. Secondly, the programmable logic controllers that are pointed to remote terminal units (RTUs) to collect data, which define the system's state. The water tank RTU ladder logic includes six setpoint registers; HH, HI, LO, and LL water level setpoint register, a pump override setpoint register, and a mode setpoint register. Furthermore, it includes three output registers which store process parameters: pump state, water level, and alarm state. Thirdly, supervisory controls are handled by the master terminal unit (MTU), which in turn forwards commands to RTU. MTU sends a read query to read from the registers to measure the state of the system. The fourth level is the human–machine interface (HMI) that is used to display the sensor data received by MTU. HMI provides an interface for an operator to monitor and control the system and operations in the form of visual representation.

HMI supports a communication protocol such as MODBUS commands. HMI (master) sends commands to MODBUS (slave) in which the individual RTU executes the command and returns a response. MTU copies commands and responses received from the HMI port to the radio port and vice versa, while HMI software makes changes (every 2 s) to setpoint register values to control the physical process [17].

9.3.2 Decision Making

In our case study, the decision tree (DT) and K-nearest neighbor (KNN) algorithms are employed as classifiers in combination with multiple dimension reduction techniques (i.e., as explained in the previous section) to enable a comprehensive study on the suitability of each DR method for the selected case study (see Fig. 9.1).

DT is a classification method that uses a representation of a tree structure consisting of internal nodes that represent a test on an attribute and branches, which denote the outcome of the test and each leaf node holds a class label. KNN is a non-parametric algorithm that uses the label information to learn new unlabeled data based on a similarity measure by calculating the distance between points using distance measures such as Euclidean distance, Hamming distance, Manhattan distance, and Minkowski distance. DT and KNN classifiers can analyze data and identify significant characteristics in the network from the IDS.

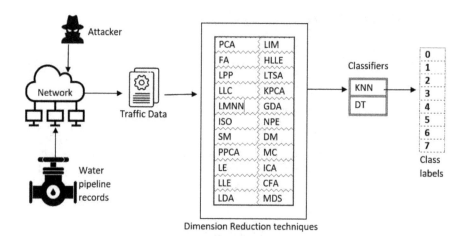

Fig. 9.1 Illustrative diagram of the designed intrusion detection system

Table 9.1 List of simulated cyber-attacks in the water storage system

Classes	Types of attacks
0	Instance not part of an injection
1	Naive malicious response injection
2	Complex malicious response injection
3	Malicious state command injection
4	Malicious parameter command injection
5	Malicious function code injection
6	Denial of service injection
7	Reconnaissance injection

The intrusion detection system detects and classifies seven different types of cyber-attacks in the water storage tank system, including a normal class when the system is safe and an injected class, as shown in Table 9.1. The network traffic data are validated and trained by incorporating state-of-the-art DR techniques into a hybrid intrusion detection system. After the traffic data is passed through dimension reduction techniques, the most relevant features are selected, and new reduced data is trained using KNN and DT classifiers. The reduced data is classified into seven different classes as well as the safe state class as shown in Table 9.1.

9.4 Experimental Results

In this section, we aim to obtain a new representation of the data, having a lower-dimensionality but with more informative features. Several experiments were performed to compare multiple DR techniques in terms of accuracy, F1-score, and standard deviation. The classification task in these experiments has been carried out using DT and kNN classifiers. We compare 22 DR methods, namely PCA, FA, CFA, MDS, LDA, ISO, SM, PPCA, LE, LLE, LIM, HLLE, LTSA, KPCA, GDA, NPE, LPP, DM, LLC, MC, LMNN, and ICA. Figure 9.1 shows the DR techniques that are utilized in the designed IDS.

9.4.1   Experiment Setting

The SCADA systems record the network flow in the water storage system, which are captured via a serial port data logger. The recorded data has 200,000 samples.

In order to minimize memory requirements and processing time, 10% of samples is randomly selected for training the computational models. The recorded network traffic data consists of 24 unique features, as shown in Table 9.2, that are used to detect malicious activities.

Table 9.2 List of parameters in water storage system

Feature name	Description
Command address	Device ID in command packet
Response address	Device ID in response packet
Command memory	Memory start position in command packet
Response memory	Memory start position in response packet
Command memory count	Number of memory bytes for R/W command
Response memory count	Number of memory bytes for R/W response
Command read function	Value of read command function code
Command write function	Value of write command function code
Response read function	Value of read response function code
Response write function	Value of write response function code
Sub-function	Value of sub-function code in the command/response
Command length	Total length of command packet
Response length	Total length of response packet
H	Value of H setpoint
HH	value of HH setpoint
L	Value of L setpoint
LL	Value of LL setpoint
Control mode	Automatic, manual, or shutdown
Control scheme	Control scheme of the water pipeline
Pump	Value of pump state
CRC rate	CRC error rate
Measurement	Water level
Time	Time interval between two packets
Result	Manual classification of the instance

The network traffic data is recoded from MODBUS traffic with RS-232 connection in which it is one byte long and each server has a unique device address. The water system devises a relief valve, a pump, alarm, meter, and a switch control scheme to maintain the water level between high and low setpoints.

An attack can be observed by the read and write commands/responses, which have a fixed length for each system. In a normal system, the error rate should be low and constant but when the system undergoes a denial-of-service attack the rates are expected to increase. If there is no error during the normal state, the response function code matches the command function. When there exists an error, the response sub-function code equals the summation of the command function code and a value of 0X80.

9.4.2   Results Analysis

We evaluated the performance of 22 dimensionality reduction methods and divided the train and test data on the basis of K-fold cross-validation approach, using $K = 10$, for each method then divided the results w.r.t. KNN and DT classifiers.

Both KPCA and GDA methods produce a small vector of dimension two with an average accuracy score between 0.90–0.96. In general, Fig. 9.2 shows the accuracy comparison of the system before and after applying dimension reduction methods, where bars show the averaged results obtained by testing on each fold. Figure 9.2a presents the accuracy before applying dimension reduction techniques on our datasets, while Fig. 9.2b shows the results after applying dimension reduction techniques. In general, decision tree classifier outperforms the KNN classifier in terms of accuracy. Considering the original dataset, KNN recorded an average accuracy of 87.894% and DT results in an average accuracy of 87.920%. Besides, when DR is applied, KNN and DT achieved 90.164% and 91.535% of average accuracy, respectively.

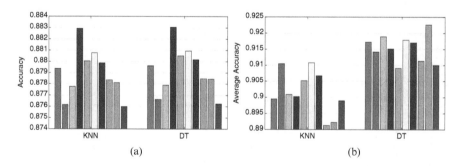

Fig. 9.2 Performance improvement in each fold of the cross-validation. (**a**) Accuracy before DR. (**b**) Accuracy after DR

In Fig. 9.3, it is apparent that performance of the PPCA method is consistently and significantly higher when combined with KNN classifier compared to other DR methods, and it reduces the dimensionality of the feature space to 10 features. In respect to DT classifier, SM method obtains the highest performance in terms of accuracy and F1-score compared to other DR methods.

Many dimensionality reduction methods perform reasonably well, and their performance is relatively stable across a range of included low-dimensional components. In terms of accuracy measure and KNN classifier, Fig. 9.3a shows that PPCA method has outperformed the other methods. This is while MC and LLC methods are ranked second and third, albeit with a slight difference. Furthermore, LDA, CFA, KPCA, ICA, LTSA, HLLE, ISO, LE, SM, LMNN, GDA, LLE, MDS, FA, LIM, DM, NPE, LPP, and PCA methods are ranked from fourth to 22-th, respectively. MC and LLC methods have desirable performance with an average accuracy from 0.97 to 0.98. LPP and PCA have failed to improve the classification performance using KNN classifier that results in an accuracy lower than 80%. Considering the results of DT accuracy in Fig. 9.3b, PPCA is ranked as first, and it is followed by LDA and MC that are ranked as second and third, with a slight difference. ISO, ICA, CFA, MDS, KPCA, LTSA, HLLE, LE, SM, GDA, FA, PCA, LMNN, LIM, LLE, and LLC methods are ranked from fourth to 19-th. DM and NPE methods share the 20-th rank as they show equal performances. Lastly, LPP was ranked as the last technique, as it recorded less than 67%.

In addition to accuracy, Fig. 9.3c, d represents the F1-score performance for KNN and DT classifiers. It also indicates that PPCA yields the highest F1-score, almost 99%, and is ranked as the best. Furthermore, MC, LLC, LDA, and CFA methods result in an average F1-score between 98% and 97%, when combined with KNN, and can be considered as the second-best algorithms. Moreover, ISO, SM, LTSA, HLLE, LLE, LE, LMNN, LIM, KPCA, MDS, and FA methods are ranked from sixth to 16-th with an average of 84–94%. On the other hand, Fig. 9.3d shows that when DT classifier is employed, the methods that come after PPCA are: LDA, MC, PCA, ISO, SM, CFA, FA, LTSA, HLLE, LE, LLE, LIM, LMNN, and LLC that are ranked from second to 15-th place. Both NPE and DM methods share the 17-th place, when KNN classifier is used, while using DT classifier they are ranked 16th. Lastly, using KNN classifier, PCA, ICA, LPP, and GDA methods are ranked as the last methods, whereas using DT classifier ICA, GDA, and LPP are ranked from 20-th to the last.

Generally, GDA, LPP, and ICA methods are not sensitive to the choice of classifiers as they result in lower F1-score than others in average of 50% and 70%. PPCA, LLC, MC, and CFA are more compatible with KNN, while the rest of the dimension reduction methods like PCA, FA, LDA, ISO, and SM are suggested to be used with DT classifier. In general, PPCA outperforms all the competitors and results in the maximum accuracy when coupled with KNN. Similarly, GDA worsens the F1-score; however, stability is improved when it is used with DT classifier.

Figure 9.4 indicates the relatively higher F1-score performance achieved by the DR methods in comparison with individual DR algorithms. The results demonstrate better performance when DT classifier is used instead of KNN classifier for most

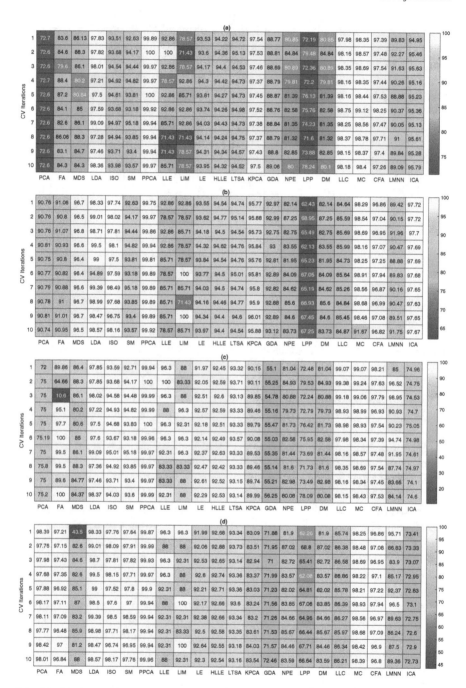

Fig. 9.3 Obtained performance measures in terms of accuracy and F1-score through 10-fold cross-validation of kNN and DT. (**a**) kNN accuracies (%). (**b**) DT accuracies (%). (**c**) kNN F1-score (%). (**d**) DT F1-score (%)

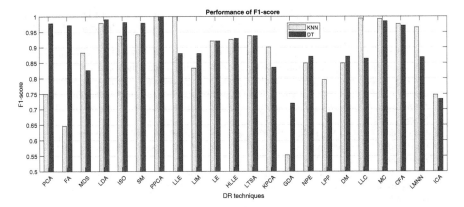

Fig. 9.4 F1-score performance in respect to DT and KNN classifiers

DR methods. The best performance in terms of the F1-score is almost 100% that is obtained by both KNN and DT classifiers in combination with PPCA and using the dimensionality size of 10 feature.

9.5   Conclusion

An intrusion detection system is prepared to study the effect of dimensionality reduction on the intrusion detection performance. A SCADA system of a water storage tank is selected as the case study. This cyber-physical system undergoes multiple cyber-attacks in our study, for which we design an intrusion detection system. The utilized IDS leverages 22 advanced dimensionality reduction techniques that are coupled with two classifiers. This hybrid scheme enables a comparative study on the impact of DR methods and their compatibility with the selected classifiers. These algorithms are compared in terms of accuracy, F1-score, and standard deviation. The conducted analysis ranks all the methods and proposes the best combination for the optimal detection accuracy in the selected case study.

References

1. Bafna, P., Shirwaikar, S., & Pramod, D. (2019). Task recommender system using semantic clustering to identify the right personnel. *VINE Journal of Information and Knowledge Management Systems*.
2. Bahmaninezhad, F., & Hansen, J. H. (2016). Generalized discriminant analysis (GDA) for improved i-vector based speaker recognition. In *Interspeech* (Vol. 2016) (pp. 3643–3647).
3. Belkin, M., & Niyogi, P. (2003). Laplacian Eigenmaps for dimensionality reduction and data representation. *Neural Computation, 15*(6), 1373–1396.

4. Buczak, A. L., & Guven, E. (2016). A survey of data mining and machine learning methods for cyber security intrusion detection. *IEEE Communications Surveys Tutorials, 18*(2), 1153–1176.
5. Chakrabarti, S., Razavi-Far, R., Saif, M., & Rueda, L. (2017). Multi-class heteroscedastic linear dimensionality reduction scheme for diagnosing process faults. In *2017 IEEE 30th Canadian Conference on Electrical and Computer Engineering (CCECE)* (pp. 1–4).
6. Donoho, D. L., & Grimes, C. (2003). Hessian eigenmaps: Locally linear embedding techniques for high-dimensional data. *Proceedings of the National Academy of Sciences, 100*(10), 5591–5596.
7. Farajzadeh-Zanjani, M., Hallaji, E., Razavi-Far, R., & Saif, M. (2021). Generative adversarial dimensionality reduction for diagnosing faults and attacks in cyber-physical systems. *Neurocomputing, 440*, 101–110.
8. Hallaji, E., Razavi-Far, R., & Saif, M. (2020). Detection of malicious SCADA communications via multi-subspace feature selection. In *International Joint Conference on Neural Networks (IJCNN)* (pp. 1–8).
9. Hammami, Z., Sayed Mouchaweh, M., Mouelhi, W., & Ben Said, L. (2018). Discussion and review of the use of neural networks to improve the flexibility of smart grids in presence of distributed renewable resources. In *2018 17th IEEE International Conference on Machine Learning and Applications (ICMLA)* (pp. 1304–1309).
10. Hammami, Z., Sayed Mouchaweh, M., Mouelhi, W., & Ben Said, L. (2020). Neural networks for online learning of non-stationary data streams: A review and application for smart grids flexibility improvement. *Artif Intelligence Review, 53*, 6111–6154.
11. He, X., Cai, D., Yan, S., & Zhang, H. J. (2005). Neighborhood preserving embedding. In *Tenth IEEE International Conference on Computer Vision (ICCV'05)* (Vols. 1, 2, pp. 1208–1213). IEEE.
12. He, X., & Niyogi, P. (2004). Locality preserving projections. In *Advances in neural information processing systems* (pp. 153–160).
13. Igure, V. M., Laughter, S. A., & Williams, R. D. (2006). Security issues in SCADA networks. *Computers & Security, 25*(7), 498–506.
14. Imperial, J. (2019). The multidimensional scaling (MDS) algorithm for dimensionality reduction. Medium-Data Driven Investor.
15. Kempfert, K. C., Wang, Y., Chen, C., & Wong, S. W. (2020). A comparison study on nonlinear dimension reduction methods with kernel variations: Visualization, optimization and classification. *Intelligent Data Analysis, 24*(2), 267–290.
16. Liao, H. J., Richard Lin, C. H., Lin, Y. C., & Tung, K. Y. (2013). Intrusion detection system: A comprehensive review. *Journal of Network and Computer Applications, 36*(1), 16–24.
17. Morris, T., Srivastava, A., Reaves, B., Gao, W., Pavurapu, K., & Reddi, R. (2011). A control system testbed to validate critical infrastructure protection concepts. *International Journal of Critical Infrastructure Protection, 4*(2), 88–103.
18. Navlani, A. (2019) Introduction to factor analysis in python.
19. Pal, A. K. (2018). Dimension reduction—isomap. Paperspace.
20. De la Porte, J., Herbst, B., Hereman, W., & Van Der Walt, S. (2008). An introduction to diffusion maps. In *Proceedings of the 19th Symposium of the Pattern Recognition Association of South Africa (PRASA 2008)*, Cape Town, South Africa (pp. 15–25).
21. Raschka, S. (2014). Linear discriminant analysis. Sebastianraschka.
22. Razavi-Far, R., Chakrabarti, S., Saif, M., & Zio, E. (2019). An integrated imputation-prediction scheme for prognostics of battery data with missing observations. *Expert Systems with Applications, 115*, 709–723.
23. Razavi-Far, R., Davilu, H., Palade, V., & Lucas, C. (2009). Model-based fault detection and isolation of a steam generator using neuro-fuzzy networks. *Neurocomputing, 72*(13), 2939–2951. Hybrid Learning Machines (HAIS 2007)/Recent Developments in Natural Computation (ICNC 2007)

24. Razavi-Far, R., Farajzadeh-Zanjani, M., Saif, M., & Chakrabarti, S. (2020). Correlation clustering imputation for diagnosing attacks and faults with missing power grid data. *IEEE Transactions on Smart Grid, 11*(2), 1453–1464.
25. Razavi-Far, R., Farajzadeh-Zanjani, M., Wang, B., Saif, M., & Chakrabarti, S. (2019). Imputation-based ensemble techniques for class imbalance learning. *IEEE Transactions on Knowledge and Data Engineering, 33*(5), 1988–2001.
26. Razavi-Far, R., Hallaji, E., Farajzadeh-Zanjani, M., & Saif, M. (2019). A semi-supervised diagnostic framework based on the surface estimation of faulty distributions. *IEEE Transactions on Industrial Informatics, 15*(3), 1277–1286.
27. Razavi-Far, R., Hallaji, E., Farajzadeh-Zanjani, M., Saif, M., Kia, S.H., Henao, H., & Capolino, G. (2019). Information fusion and semi-supervised deep learning scheme for diagnosing gear faults in induction machine systems. *IEEE Transactions on Industrial Electronics, 66*(8), 6331–6342.
28. Razavi-Far, R., Hallaji, E., Saif, M., & Ditzler, G. (2019). A novelty detector and extreme verification latency model for nonstationary environments. *IEEE Transactions on Industrial Electronics, 66*(1), 561–570.
29. Razavi-Far, R., Palade, V., & Zio, E. (2014). Optimal detection of new classes of faults by an invasive weed optimization method. In *International Joint Conference on Neural Networks (IJCNN)* (pp. 91–98).
30. Rosman, G., Bronstein, M. M., Bronstein, A. M., & Kimmel, R. (2010). Nonlinear dimensionality reduction by topologically constrained isometric embedding. *International Journal of Computer Vision, 89*(1), 56–68.
31. Sayed-Mouchaweh, M. (Ed.) (2018). Diagnosability, *Security and safety of hybrid dynamic and cyber-physical systems*, 1st ed. Springer.
32. Solutions, S. (2013). Confirmatory factor analysis. Retrieved May 28, 2016.
33. Sumithra, V., & Surendran, S. (2015). A review of various linear and non linear dimensionality reduction techniques. *International Journal of Computer Science and Information Technologies, 6*, 2354–2360.
34. Sun, S., & Chen, Q. (2011). Hierarchical distance metric learning for large margin nearest neighbor classification. *International Journal of Pattern Recognition and Artificial Intelligence, 25*(7), 1073–1087.
35. Tipping, M. E., & Bishop, C. M. (1999). Probabilistic principal component analysis. *Journal of the Royal Statistical Society: Series B (Statistical Methodology), 61*(3), 611–622.
36. Van Der Maaten, L., Postma, E., & Van den Herik, J. (2009). Dimensionality reduction: A comparative. *Journal of Machine Learning Research, 10*(66–71), 13.
37. Zhang, Z., & Zha, H. (2002). Principal manifolds and nonlinear dimension reduction via local tangent space alignment. *SIAM Journal of Scientific Computing, 26*, 313–338.

Index

Printed in the USA
CPSIA information can be obtained
at www.ICGtesting.com
LVHW082254041123
762998LV00005B/375

9 783030 764081